D1113140

BUILDER'S PORTABLE HANDBOOK

BUILDER'S PORTABLE HANDBOOK

AUGUST W. DOMEL, JR., PH.D., J.D.

McGraw-Hill

New York San Francisco Washington, D.C. Auckland Bogotá
Caracas Lisbon London Madrid Mexico City Milan
Montreal New Delhi San Juan Singapore
Sydney Tokyo Toronto

Library of Congress Cataloging-in-Publication Data

Domel, August W. (August William),
    Builder's portable handbook / August W. Domel, Jr.
        p.   cm
    ISBN 0-07-134654-6
    1. Building—Handbooks, manuals, etc.   I. Title.
    TH151 .D56 2000
    690—dc21   00-022008

McGraw-Hill

A Division of The McGraw·Hill Companies

1 2 3 4 5 6 7 8 9 0 DOC/DOC 0 9 8 7 6 5 4 3 2 1 0

ISBN 0-07-134654-6

This book was set in Times by the author.

Printed and bound by R. R. Donnelley and Sons Co.

McGraw-Hill books are available at special quantity discounts to use as premiums and sales promotions, or for use in corporate training programs. For more information, please write the Director of Special Sales, McGraw-Hill, Two Penn Plaza, New York, NY 10121-2298. Or contact your local bookstore.

This book is printed on recycled, acid-free paper containing a minimum of 50% recycled, de-inked fiber.

Contents

Disclaimer

This text was written for the purpose of introducing contractors and other individuals to the technical side of residential construction. Reading this text by no means makes one a qualified designer, engineer or architect. It will however make the reader better informed to participate in the design process, fixing field problems and understanding the technical side of their work product.

The information contained herein was written specifically for single-family residential construction and light construction. Discussions, calculations, figures, and tables were prepared with this structure in mind. It is not advisable that the principles in this text be used for any other type of structure.

This text was reviewed for accuracy and for technical content. A great deal of effort was spent on trying to produce an error-free text. In reality, such a goal is unattainable. Therefore, I caution the reader of the possibility of mistakes in the book. I encourage that the reader notify the publisher if any mistakes are found. The author assumes no responsibility for anyone's use of this text and its principles.

Dedication

This book is dedicated to my wife Gina, my daughter Tommasina, and my sons Gus, Antonio, Gino, and Reno.

Acknowledgments

The author sincerely acknowledges the contributions of Lisa Lehocky and Zoe G. Foundotos. Lisa Lehocky provided the desktop publishing for this entire book, her patience during the numerous rounds of editing was greatly appreciated. Guidance and suggestions provided by Zoe Foundotos were essential in achieving the quality of this text.

Chapter 1
Introduction

CHAPTER 1—INTRODUCTION

The purpose of this book is to provide the reader with a reference source for the rapid retrieval or a quick introduction of technical and non-technical information for residential and other light construction. The book was designed to be utilized as a handbook to be consulted when the reader's personal library could not be accessed.

The topics were chosen based on the author's field experience. Many of the topics were included because the author at one time had a need for this information in the field but a resource was not immediately available.

The audience that will find this book useful is primarily the residential and light commercial contractor. The book should assist in resolving common field questions, allowing the project to forge onward. It was also written with the architect, engineer and building official in mind. These professionals will find the many topics presented in a compact format to be of use when in the field.

Much of this book is presented in a tabular or graphic format. This was done to achieve the goal of providing a resource that would supply quick and concise answers. Excessive amounts of text were avoided since it would frustrate the goal of it being a quick reference.

It is suggested that the reader browse through the book from front cover to back cover to know the topics that are presented herein. This will allow the reader to become familiar enough with the book to use it effectively when necessary.

The book has 12 chapters beyond this introduction. A brief overview of each chapter follows.

CHAPTER 2—SAFETY

The fastest way to detour a profitable and successful project is to have an injury occur on the jobsite. The corollary to this is that a safe work site is a successful work site.

Preparation for action for a jobsite injury can mitigate the damage of the injury. It is useful to have some minimal equipment for use in the event of an injury, including band-aids or a clean cloth. Also, in areas where 911 emergency response is not available, a plan for response to a serious accident is needed.

This chapter presents some of the OSHA regulations that may apply to residential or light commercial construction projects. Issues covered include material handling, excavation, demolition and other topics.

CHAPTER 3—FOUNDATIONS

Excavation can be a venture into the unknown. On larger projects, resources can be expended on determining the soil profile and other useful information of the subsurface conditions. It is often not feasible to allocate resources to soil investigation on a typical smaller project. This can result in the encountering of unknown, underground hidden conditions including boulders, unsuitable soils and a high water table.

Soil behavior is directly related to its permeability characteristics, particle size, strength capacity and other characteristics. The type and extent of soil sampling needed depends on the structure to be built. If a septic system is to be used, the drainage characteristics and the profile of the soil must be evaluated. When considering whether a soil can support a vertical load, the load bearing capacity of the soil must be determined as well as the soil's resistance to settlement.

This chapter presents information on soil characteristics and how soils are classified. Foundation construction details are provided for dampproofing, fill heights and general dimension requirements.

CHAPTER 4—LOADS

Building loads can be divided into two types: vertical and horizontal. Vertical loads, those produced by gravity, consist of dead loads (self-weight) and live loads (weight of non-permanent loads). These loads can be fairly well predicted for residential and light construction.

Lateral loads result from wind and seismic forces, although these can also produce vertical loads. These loads are much more difficult to predict than other loads.

Building codes provide detailed information on the magnitude and location of loads. Distribution of the load is as important as the magnitude. A 1000 pound load distributed over 10 feet provides different stresses than the same load spread over one foot. Careful consideration should be given to all potential loading conditions and their loading distribution.

Economic considerations often control the size of a structure. Economic constraints typically lesson overtime and a building expansion may be an option. The expenses of an expansion can be reduced if the possibility of expansion is considered while determining the loads.

This chapter presents the building code prescribed loads for live and dead loads. Discussions are also presented on wind and seismic loads.

CHAPTER 5—CONCRETE

Concrete is an excellent choice of material for all size construction projects. It can be formed into a variety of shapes that dry into a hard, durable product. The negative side of concrete is that to have an aesthetically pleasing product, you must have proper placement techniques.

Proper placement includes having a concrete mix that will have sufficient strength but yet be liquid enough to be placed. This can only be achieved by proper planning for the mix design and for field placement.

This chapter provides the basics of the technical aspects of concrete construction. Field information such as curing, tolerances and testing are presented. Technical aspects presented include code requirements, mix design and terminology.

CHAPTER 6—MASONRY

Masonry serves a dual function in construction. It provides structural integrity to the building as well as an architectural finish. Concrete block can be used for construction of foundation walls or walls above grade. Clay brick can be used for the exterior facade over concrete block or as veneer for wood framing. A variety of brick patterns can be used in conjunction with different mortar joint preparations to provide an architectural finish.

Masonry construction requires that modular tolerances be followed to allow whole bricks to be used. Mortar joints must be installed properly and expansion allowance for heat and cold movement must be used to prevent cracking.

This chapter presents architectural features of masonry including arch typing, joint finishes and brick wall patterns. Construction information such as tolerances and mortar types is also provided.

CHAPTER 7—WOOD

Wood has a multitude of use in residential and light construction. It is used for structural members for foundations, walls, floors and roofs. It is used on the exterior of a structure for cladding in the form of decking and siding and on the interior for trim and doors.

Wood is unique in that it does not have uniform properties in one direction as compared to any other direction. This is because wood has different properties in relation to where it is removed from the tree. Also, the strength of wood depends on its moisture content, the presence of knots and the nature of the loads.

This chapter presents tabular information for selecting wood joists and rafters and for the bracing of wood trusses. Information is also provided on material properties, geometric properties and choices of millwork.

CHAPTER 8—STEEL

Steel is for the most part relegated to use as beams and columns located in the basement or crawl space. Beams are typically 8 or 10 inches deep. Steel columns are usually pipes that are 3 1/2 inches in diameter and filled with concrete.

It is prudent to consider ceiling clearance at the location of the steel beam since there will be an 8 to 10 inch drop. The HVAC system can parallel the beam and be included in the same soffit as the steel beam.

Steel beams come in standard sizes and shapes. It is worthwhile to understand the basis of the designations. This knowledge comes in use when size substitutions are necessary.

This section presents information on geometric properties of I-shaped beams and angle-shaped members. Tables are also provided to assist in the selection of beams and lintels.

CHAPTER 9—GEOGRAPHICAL DATA

This chapter provides information on conditions in the 48 contiguous states. Information is provided for wind speeds, seismic potential, temperature ranges and similar data.

CHAPTERS 10 AND 11—RESOURCES

These chapters provide sources of information for contacting suppliers and technical resources, respectively.

CHAPTER 12—MISCELLANEOUS

Information is presented on roofing membranes, septic disposal systems, geometric properties of various shapes and conversion factors.

Chapter 2
Safety

TIPS & INFORMATION

- Before starting a construction project, know where emergency medical facilities are located and how to contact local emergency services.

- A first aid kit that includes bandages and a clean cloth comes in handy for minor injuries.

- OSHA regulations are prepared by the United States Government Department of Labor. Your local OSHA office can provide you with a copy of OSHA construction safety requirements.

- Local construction safety councils can provide a wealth of information and also provide safety training courses.

- Manufacturers of materials provide safety data sheets for their products. These sheets provide detailed information on hazards of products.

- A structure under construction is often unstable. Precautions must be taken to temporary brace components that are unstable.

- Accidents can occur from having a sloppy work site. A messy work site not only presents tripping hazards, it sets a bad tone for the project in general.

- Avoid storing materials in concentrated areas of the structure. A roof structure may be incapable of supporting the weight of roof shingles piled in one location. This condition is even more critical when material is stored on an incompleted structure.

This chapter presents safety considerations for light construction. The requirements presented are based on those provided in the Occupational Safety and Health Standards for the Construction Industry, which is better known as OSHA.

Information presented herein is by no means the minimum or maximum of safety requirements that should be known and enforced by the contractor. The information is provided to present an overview of some of the basic safety requirements.

This text was written for residential and light construction. The OSHA regulations contain nearly 600 pages of requirements for the construction industry. Only a very small portion of those require-ments are presented here. If the project being worked on is unusual or larger than light construction, there would most likely be many other requirements that would need to be considered. This chapter provides very general safety information. No information contained herein should be used for the purpose of showing what entity is responsible for any particular portion of safety, that a particular task must be performed in any certain matter, that a certain device is inappropriate for a certain task, etc. The use of the safety informa-tion contained herein for determining contractual obligations, who is responsible for safety, or if a construction procedure is safe is beyond the intent of the author and an inappropriate use.

GENERAL SAFETY INFORMATION

- Debris shall be kept clear from work areas and other areas where they pose a danger throughout the construction project.

- Natural or artificial illumination shall be provided in construction areas, aisles, stairs, ramps, etc.

PERSONAL PROTECTIVE EQUIPMENT

- Employer shall be responsible to assure adequacy, proper mainte-nance and sanitation of employee provided protective equipment.

- Protective helmets shall be used by employees working in areas where there is possible danger of head injury from impact or falling objects.

- Employees shall be provided with eye and face protection equipment when machines or operations present the potential for eye or face injury.

MATERIAL HANDLING AND RELATED ISSUES

- All materials stored in tiers shall be secured to prevent sliding, falling or collapse.

- Aisles and passageways shall be kept clear to provide for safe movement of material and employees.

- Material stored inside buildings under construction shall not be placed within six feet of an inside floor opening or within ten feet of an exterior wall that does not extend above the pile of material.

- Brick stacks shall not be more than seven feet high. When a brick stack reaches four feet it shall be tapered back two inches for each additional foot.

- Used lumber shall have all nails removed before stacking.

- Lumber to be handled manually shall not be stacked more than 16 feet high.

- Rigging equipment shall be inspected prior to its use on each sift.

- Chain slings shall have permanently affixed durable identification stating size, grade, rated capacity and sling manufacturer.

- Wire rope shall not be used if the visible broken wires exceed 10 percent of the total number of wires in the following lengths of wire:

    1/4" diameter → 2" of rope length
    3/8" diameter → 3" of rope length
    1/2" diameter → 4" of rope length
    5/8" diameter → 5" of rope length

RATED CAPACITY FOR SINGLE CHAIN VERTICAL LIFT

1/4" Chain	3250 pounds
3/8" Chain	6,600 pounds
1/2" Chain	11,250 pounds
5/8" Chain	16,500 pounds

RATED CAPACITIES IN POUNDS FOR SINGLE LEG WIRE ROPE SLINGS FOR VERTICAL LIFT (6 × 19 WIRE CONSTRUCTION)

	Hand Spliced	Swaged
1/4	1060	1180
3/8	2200	2600
1/2	4000	4600
5/8	6000	7200

RATED CAPACITIES IN POUNDS FOR SINGLE LEG WIRE ROPE SLINGS FOR CHOKER LIFT (6 × 19 WIRE CONSTRUCTION)

	Hand Spliced	Swaged
1/4	800	880
3/8	1720	1960
1/2	3000	3400
5/8	4400	4700

- Whenever materials are dropped more than 20 feet to any point outside the exterior walls, an enclosed chute shall be used.

TOOLS

- All hand tools and power tools furnished by the employer or employee shall be maintained in a safe condition.

- Employers shall not issue or permit the use of unsafe hand tools.

- Wooden handles of tools shall be kept free of splinters or cracks.

- Electric cords shall not be used to hoist or lower tools.

- Compressed air shall not be used for cleaning except where reduced to less than 30 psi.

SCAFFOLDS

- The front edge of platforms shall be no more than 14 inches from the face of the work. For plastering and lathing operations, this distance shall be no more than 18 inches.

- Scaffolds with a height to base width of greater than four to one shall be restrained from tipping.

- Footings shall be level, sound, rigid and capable of supporting the scaffold load.

- Scaffolds shall not be moved horizontally while employers are on them unless designed for such use.

- Each employee on a scaffold more than 10 feet above a lower level shall be protected from falling to a lower level.

- The top edge height of top rails of guard rail systems shall be installed 38 inches to 45 inches above the platform surface.

FALL PROTECTION

- Each employee on a working surface with an unprotected edge shall be provided with fall protection.

- Body belts are not acceptable as a personal fall arrest system.

EXCAVATIONS

- A stairway, ladder, ramp or other safe means of egress shall be located in trench excavations that are four feet or more in depth.

- Utility companies or owners shall be contacted within established or customary local response time, advised of the proposed work and asked to establish locations.

- When utilities cannot respond within 24 hours or a larger period as required by government officials or cannot locate the utility, the employer may proceed with caution and with an acceptable means to locate the buried utility.

- When mobile equipment is operated adjacent to an excavation and operator does not have a clear and direct view of the edge, barricades, signals or stooping devices shall be used as warning devices.

CONCRETE AND MASONRY CONSTRUCTION

- All protruding reinforcing steel shall be guarded to eliminate the hazard of impalement.

- No employee shall be permitted to work under a concrete bucket while buckets are being lifted or lowered.

- A limited access zone shall be established prior to starting construction of a masonry wall.

- The limited access zone shall be equal to the height of the wall plus four feet for the entire length of the wall.

- The limited access zone shall be established on the side for which there will be no scaffolding.

- The limited access zone shall be restricted to employees actively engaged in construction of the wall. No other employees shall be permitted to enter the zone.

DEMOLITION

- Prior to permitting the employees to start demolition operations, an engineering survey shall be made by a competent person to determine the condition of the structure and possibility of unplanned collapse.

- When employees are required to work within a structure to be demolished which has been damaged by fire, flood, explosion or other cause, the walls or floors shall be braced or shored.

- All utility service lines shall be shut off before demolition work begins.

- It shall be determined if any hazardous material is present and appropriate identification testing shall be performed.

- Walkways or ladders shall be provided to enable employees to safely reach or leave any scaffold or wall.

STAIRWAYS AND LADDERS

- Stairs shall be installed between 30 degrees and 50 degrees from the ground surface.

- A stairway or ladder shall be provided at access locations where there is a break in elevation of 19 inches or more.

- Stairways more than 30 inches from grade or having four or more risers shall have one handrail.

- Riser height and thread depth shall be uniform within each flight of stairs.

- Rungs of ladders shall be spaced not less than 10 inches apart nor more than 14 inches apart.

- The minimum clear distance between side rails for all portable ladders shall be 11 1/2 inches.

- Wood ladders shall not be coated with any opaque covering.

NOTES:

NOTES:

NOTES:

NOTES:

Chapter 3
Foundations

TIPS & INFORMATION

- Soil testing prior to purchasing a site can avert a future disaster. For a relatively small fee, soils testing can be performed to identify soils susceptible to shrinkage and swelling. Also, if applicable, the suitability of the site to support a septic system can be evaluated.

- It is incumbent on the contractor to contact the board of underground prior to excavating in areas where utilities may be buried. Always proceed with caution: the locations marked for utilities may not always be accurate.

- An excavation that is deeper than the waist has the potential to collapse. Sloping back the sidewalls of an excavation is a simple solution to a dangerous problem.

- The footing is the first structural element installed. Check the elevation and location of the footing prior to proceeding with other work. A footing that is not level will require that other trades adjust their work.

- Address water infiltration problems early in the project. Excessive water in the excavation pit or percolating at isolated locations could be a sign that a second sump pit may be needed.

- Backfilling prior to installation of the first floor framing makes access easier for the carpenters, but creates the possibility of failure of the wall. A rain fall can saturate the soil, greatly increasing its lateral pressure. Backfill only after bracing the wall or determining it is capable of resisting the applied soil pressure.

- The soil grade should slope away from the foundation. Ground with no slope allows for water to collect at the wall and increase the likelihood of water infiltration.

- Check with local geological societies when constructing above areas where coal mining has occurred. Consider building elsewhere or design for potential ground movement.

- A properly sized keyway is needed at the interface of the wall and footing. A 2 × 4 on edge inserted into the footing during initial set is a good way to provide a properly formed keyway. A line scribed by a reinforcing bar is virtually useless.

- Never place concrete on a base that is not compacted. Placing the base early and allowing it to initially compact under environmental conditions is a good start.

DEFINITIONS

Bedrock	Rock located in its place of origin usually underneath soils of various depths. Solid, and a desirable material for supporting foundations with heavy loads.
Boulders	Large pieces of material that were once part of bedrock. The sizes of boulders have a dimension larger than one foot and can be problematic when encountered during excavation work.
Clay	Mineral material that is smaller than silt particles. Has high cohesion capabilities. Has varied range of support capacity.
Glacial Till	A mixture of soil material that was picked up and deposited by glaciers.
Gravel	Pieces of rock that have a dimension no greater than six inches and no less than 1/4 inch. This material is often placed as a base material for foundations, slabs and roads.
Hardpan	Term commonly used for firm soils.
Loam	Material that contains organics.
Peat	Organic material that is partly decayed.
Sand	Mineral material that is smaller than 1/400 of an inch. Sand does not stick (i.e., is cohesionless) to adjacent sand particles.
Shale	Rock composed of clay material compressed together.
Silt	Mineral material that is smaller than sand particles but larger than 1/13,000th of an inch. Has some limited cohesion.

SOIL SIZES AND STRENGTHS

CLAY

•Size:	1/13,000" to 1/25,000"
•Bearing Capacity:	Soft Clays–500 psi to 2000 psi
	Medium Clays–2000 psi to 3500 psi
	Stiff Clays–3500 psi to 8000 psi

SAND

•Size:	1/4" to 1/400"
•Bearing Capacity:	Loose–1500 psi to 4000 psi
	Compact–2500 psi to 8000 psi

SILT

•Size:	1/400" to 1/13,000"
•Bearing Capacity:	2000 psi to 3000 psi

GRAVEL

•Size:	6" to 1/4"
•Bearing Capacity:	8000 psi to 20,000 psi

ROCK

•Bearing Capacity:	12,000 psi to 200,000 psi

Table 3.1—Soil Classification
ASTM Classification System

Symbol	Gravel	Sand	Silt	Clay	Peat
GW	Yes–well graded	Yes	—	—	—
GP	Yes–poorly graded	Yes	—	—	—
GM	Yes	Yes	Yes	—	—
GC	Yes	Yes	—	Yes	—
SW	Yes	Yes–well graded	—	—	—
SP	Yes	Yes–poorly graded	—	—	—
SM	—	Yes	Yes	—	—
SC	—	Yes		Yes	—
ML	—	—	Inorganic	—	—
CL	—	—	—	Inorganic	—
OL	—	—	Organic	—	—
MH	—	—	Inorganic – fine grain	—	—
CH	—	—	—	Inorganic – highly plastic	—
OH	—	—	—	Organic	—
Pt	—	—	—	—	Yes

OSHA SOIL CLASSIFICATIONS

TYPE A SOIL

Consistency: Cohesive

Strength: Greater than 3000 pounds per square foot

Examples: Subject to heavy construction traffic or pile
                  driving.
                  Fissures are present.
                  Soil was previously disturbed.

TYPE B SOIL

Consistency: Cohesive and Cohesionless

Strength: 1000 to 3000 pounds per square foot

Examples: Clays, Silty Clays, Sandy Clays, Crushed Rock,
                  Silt, Silty Loam and Sandy Loam.

Includes: Type A soil that has been previously disturbed.
                  Type A soil that has fissures.
                  Type A soil near heavy construction traffic or pile
                  driving.

TYPE C SOIL

Consistency: Cohesive and Cohesionless

Strength: Less than 1000 pounds per square foot

Examples: Many types including sand and gravel.

Includes: Submerged soil and solid which has freely seeping
                  water.

SIMPLE SLOPE EXCAVATIONS PER OSHA

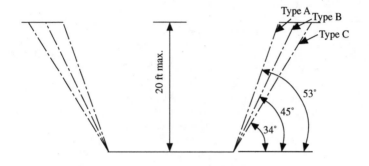

SLOPE WITH BENCH PER OSHA

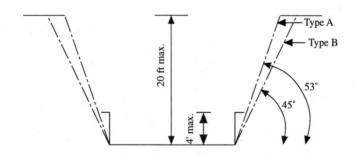

Figure 3.1—Excavation Side Slopes

SHORT TERM SIMPLE SLOPE PER OSHA

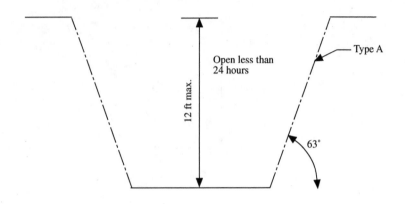

SHALLOW EXCAVATION PER OSHA

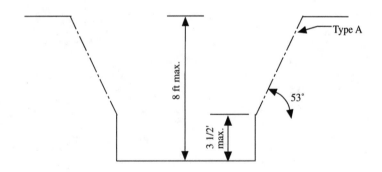

Figure 3.1—(cont.) Excavation Side Slopes

SHIELD EXCAVATION

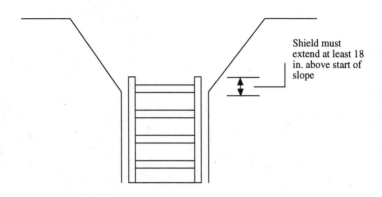

Shield must
extend at least 18
in. above start of
slope

LAYERED SOILS

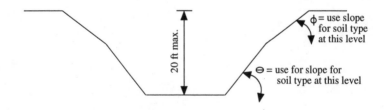

20 ft max.

φ = use slope
for soil type
at this level

θ = use for slope for
soil type at this level

Figure 3.1—(cont.) Excavation Side Slopes

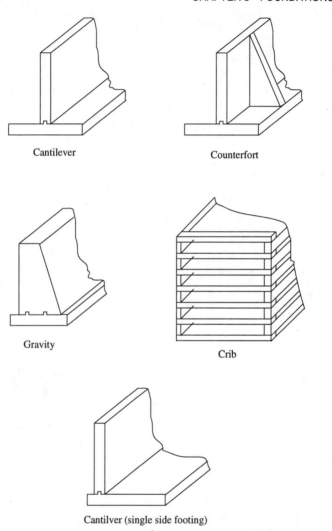

Cantilever

Counterfort

Gravity

Crib

Cantilver (single side footing)

Figure 3.2—Retaining Wall Types

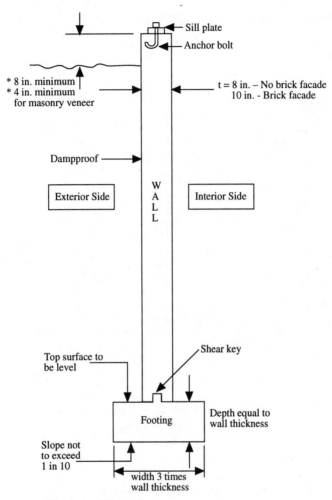

Figure 3.3—Typical Foundation Wall–Concrete
Subject to Typical Conditions

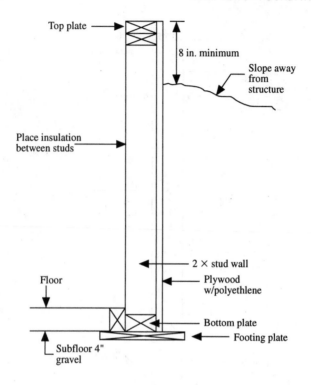

Figure 3.4—Foundation Wall–Wood
Subject to Typical Conditions

- Stud spacing 12 or 16 inches

- For 12 inch stud spacing, use 1/2 inch plywood

- For 16 inch stud spacing, use 3/4 inch plywood

- If wall is greater than six feet, use 3/4 inch plywood

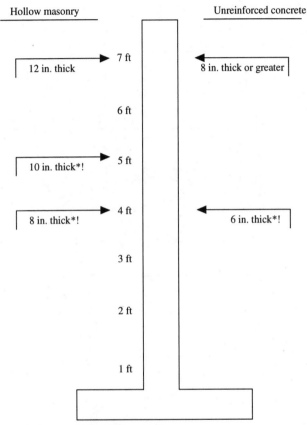

Hollow masonry

Unreinforced concrete

12 in. thick → 7 ft ← 8 in. thick or greater

6 ft

10 in. thick*! → 5 ft

8 in. thick*! → 4 ft ← 6 in. thick*!

3 ft

2 ft

1 ft

*May be increased 1 foot for better soils
! Slight increase allowed when facade is masonry

Figure 3.5—Allowable Depth of Unbalanced Fill
Typical Non-Seismic Condition

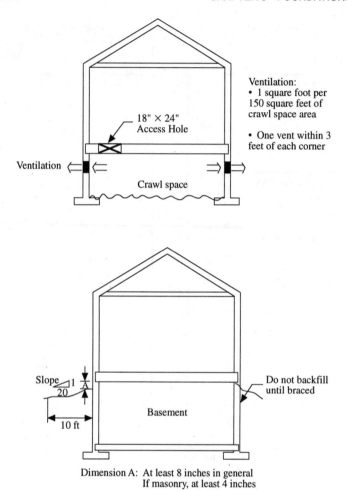

Ventilation:
• 1 square foot per 150 square feet of crawl space area

• One vent within 3 feet of each corner

18" × 24" Access Hole

Ventilation

Crawl space

Slope 1/20

Do not backfill until braced

Basement

10 ft

Dimension A: At least 8 inches in general
If masonry, at least 4 inches

Figure 3.6—General Foundation Information

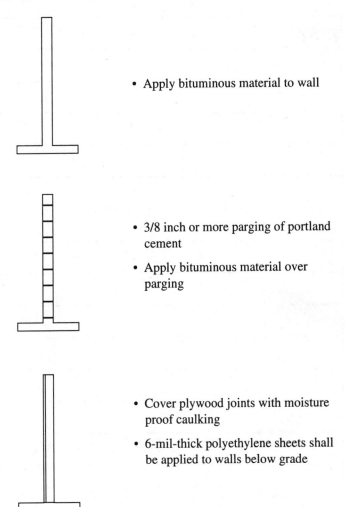

- Apply bituminous material to wall

- 3/8 inch or more parging of portland cement
- Apply bituminous material over parging

- Cover plywood joints with moisture proof caulking
- 6-mil-thick polyethylene sheets shall be applied to walls below grade

Figure 3.7—Damp Proofing Foundation

NOTES:

NOTES:

NOTES:

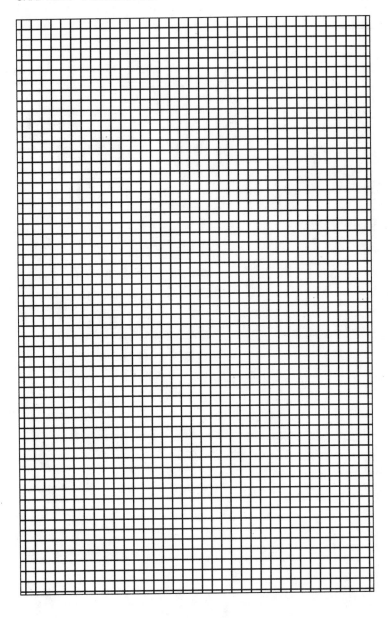

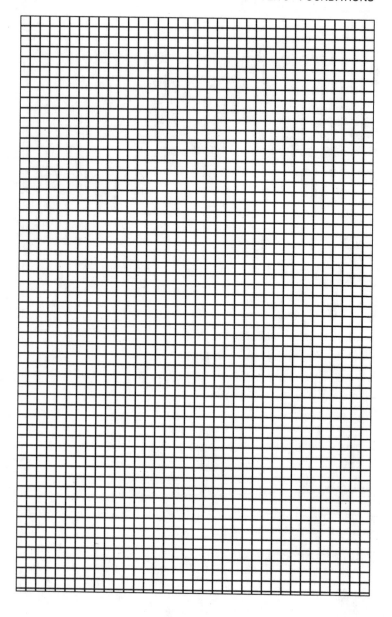

Chapter 4
Loads, Forces & Deflections

TOPICS
- Material Weight
- Live Load
- Application of Snow Load
- Lateral Foundation Load
- Wind Load
- Seismic Load

TIPS & INFORMATION

- Be generous when considering the weight of materials. Overestimating the load provides a factor of safety for changed field conditions.

- Many structural members for residential construction are controlled by serviceability requirements, not by stress. A 2×10 floor joist may be adequate to support the loads but may bounce during use, necessitating a 2×12 joist.

- A 500 pound load concentrated in a once square foot area is more critical than a 500 pound load distributed over a two foot square area. It is important to consider the area of the applied as well as the magnitude of the load.

- Always consider the ramifications of building adjacent structures of differing heights. This condition can allow snow to drift on the lower roof, producing snow loads several times of what normally may occur.

- Saturated soils push significantly more in a lateral direction as compared to drained soils. Do not allow water to accumulate at unbraced or unfinished foundation walls.

- Building codes only require that buildings be designed for a fraction of the earthquake force that may occur. This theory provides that buildings will be heavily damaged but not collapse when subjected to earthquake forces.

- Wind forces can tear off portions of the roof or wall. This makes the structure vulnerable to rain infiltration, which can cause damage that exceeded the actual wind damage. Care should be taken during construction to properly connect members to each other.

- Always consider that additional loads may be placed on the structure in the years to come.

- Consider future expansion of the structure. If a second floor addition is foreseeable, it may be wise to allow for this situation when installing the foundation.

- Be selective when choosing a material for backfill. Different soils produce differing lateral loads.

DEAD LOADS

Table 4.1—Construction Elements

Item	PCF	kN/m³
Aluminum	170	25
Bronze	550	82
Cement	90	13
Glass	160	24
Regular Concrete	145	22
Lightweight Concrete	105	16
Reinforced Concrete	150	22
Masonry (Brick)	130	19
Masonry (Stone)	165	25
Mortar	130	19
Particleboard	45	7
Plywood	36	5
Sand	100	15
Steel	490	73
Terra Cotta	120	18
Wood (Light)	30	5
Wood (Heavy)	50	7

Sample Calculation: Weight of 10 foot long 2×4 heavy lumber

Length	$= 10$ feet
Width	$= 1.5$ inches $= 0.125$ feet
Height	$= 3.5$ inches $= 0.292$ feet
Volume	$= 10 \times 0.125 \times 0.292 = 0.365$
Weight	$= 0.365 \times 50$ pcf $= 18.3$ pounds

Table 4.2—Framing Elements

Item	psf	kN/m²
2 × 4 wood stud wall	4	0.19
2 × 4 wood stud wall (1 side drywall)	8	0.38
2 × 4 wood stud wall (2 side drywall)	12	0.58
Stud walls with brick veneer	50	2.40
2 × 6 joists with flooring (16" o.c.)	5	0.24
2 × 8 joists with flooring (16" o.c.)	6	0.29
2 × 10 joists with flooring (16" o.c.)	6	0.29
2 × 12 joists with flooring (16" o.c.)	7	0.34
Ungrouted masonry block (8" thick)	40	1.92
Ungrouted masonry block (10" thick)	47	2.26
Ungrouted masonry block (12" thick)	54	2.59
Fully grouted masonry (8" thick)	85	4.08
Fully grouted masonry (10" thick)	105	5.04
Fully grouted masonry (12" thick)	125	6.00

Sample Calculation: Weight of 8 foot high stud wall, 10 foot long with two sided drywall

Height	= 8 feet
Length	= 10 feet
Area	= 8 feet × 10 feet = 80 square feet
Weight	= 80 square feet × 12 psf
	= 960 pounds

Table 4.3—Architectural Elements

Use	Item	psf	kN/m^2
Ceiling	Plaster on wood lathe	8	0.38
Ceiling	Acoustical fiber board	1	0.05
Ceiling	1/2 in. gypsum board	2.5	0.12
Roofing	Asphalt shingles	2	0.10
Roofing	4 ply felt and gravel	6	0.26
Roofing	Metal decking	3	0.14
Roofing	Insulation	1 to 1.5	0.001 to 0.0015
Roofing	Slate	10	0.48
Roofing	Wood shingles	3	0.14
Roofing	Single ply membrane	1	0.04
Flooring	Hardwood	4	0.19
Flooring	Linoleum	1	0.05
Flooring	Terrazzo	20	0.10
Flooring	Subflooring	3	0.14

Sample Calculation: Weight of linoleum and subflooring for 10 foot by 12 foot area

Length	= 10 feet
Width	= 10 feet
Area	= 100 square feet
Load	= 1 psf (linoleum) + 3 psf (subfloor) = 4 psf
Weight	= 100 $\times$ 4 = 400 pounds

LIVE LOADS

Table 4.4—Residential Single and Multifamily

Type of Structure	Location	psf	kN/m^2
One or two family	All locations unless noted otherwise	40	1.9
One or two family	Sleeping areas	30	1.4
One or two family	Habitable attics	30	1.4
One or two family	Uninhabitable attics with storage	20	1.0
One or two family	Uninhabitable attics without storage	10	0.4
One or two family	Balconies less than 100 square feet	60	2.9
Multifamily	Private rooms	40	1.9
Multifamily	Corridors serving private rooms	40	1.9
Multifamily	Public rooms	100	4.8
Multifamily	Corridors serving public rooms	100	4.8
Multifamily	Balconies	100	4.8
One family	Firescape	40	1.9

* Apartments with more than two dwellings shall be classified as multifamily.

* Hotels shall be classified as multifamily.

Table 4.5—Various Structures

Type of Structure	Location	psf	kN/m^2
Bowling Alleys	All	75	3.6
Hospitals	Private room	60	2.9
Hospitals	Operating room	40	1.9
Hospitals	Wards	40	1.9
Hospitals	Corridors above first floor	80	3.8
Libraries	Reading rooms	60	3.6
Libraries	Stack rooms	150	7.2
Office Buildings	Lobby	50	2.4
Office Buildings	Office	80	3.8
Schools	Classroom	40	1.9
Schools	First floor corridor	80	3.8
Schools	Corridors above first floor	100	4.8
Retail Stores	First floor	100	4.8
Retail Stores	Above first floor	75	3.6
Wholesale Stores	All floors	125	8.6

ROOF LIVE AND SNOW LOAD

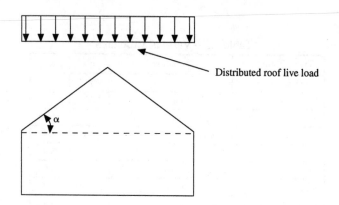

Distributed roof live load

Table 4.6—Reduced Live Load

Roof Slope	Reduced Roof Live Load Percentage	Reduced Snow Loads
15	100	100
20	100	100
25	80	100
30	80	100
35	80	88
40	80	75
45	60	63
50	60	50
55	60	37
60	60	25
65	60	13
70	60	0

* Assuming roof live load is 20 psf.

SNOW LOAD DRIFTING ON LOWER ROOF

Table 3.7—Height of Snow Drift

Ground Snow Load (psf) / Width of Building (ft)	10	15	20	25	30	35	40
25	1.2	1.3	1.4	1.6	1.6	1.8	1.8
50	1.8	2.0	2.2	2.3	2.5	2.5	2.7
75	2.3	2.6	2.7	2.9	3.0	3.1	3.3
100	2.7	3.0	3.1	3.3	3.5	3.6	3.7
125	3.0	3.2	3.4	3.7	3.9	4.0	4.1

Table 3.8—Width of Snow Drift

Ground Snow Load (psf) / Width of Building (ft)	10	15	20	25	30	35	40
25	4.8	5.2	5.6	6.4	6.4	7.2	7.2
50	7.2	8.0	8.8	9.2	10.0	10.0	10.8
75	9.2	10.4	10.8	11.6	12.0	12.4	13.2
100	10.8	12.0	12.4	13.2	14.0	14.4	14.8
125	12.0	12.8	13.6	14.8	15.6	16.0	16.4

SNOW LOAD DRIFTING AND SLIDING
ON TO LOWER ROOF

Table 3.9—Height of Snow Drift

Ground Snow Load (psf) / Width of Taller Building (ft)	10	15	20	25	30	35	40
25	1.7	1.8	2.0	2.2	2.2	2.5	2.5
50	2.5	2.8	3.1	3.2	3.5	3.5	3.8
75	3.2	3.6	3.8	4.1	4.2	4.3	4.6
100	3.8	4.2	4.3	4.6	4.9	5.0	5.2
125	4.2	4.5	4.8	5.2	5.5	5.6	5.7

Table 4.10—Width of Snow Drift

Ground Snow Load (psf) / Width of Taller Building (ft)	10	15	20	25	30	35	40
25	4.8	5.2	5.6	6.4	6.4	7.2	7.2
50	7.2	8.0	8.8	9.2	10.0	10.0	10.8
75	9.2	10.4	10.8	11.6	12.0	12.4	13.2
100	10.8	12.0	12.4	13.2	14.0	14.4	14.8
125	12.0	12.8	13.6	14.8	15.6	16.0	16.4

LATERAL LOADS ON FOUNDATION WALLS

When soil is piled up against a wall it will push against the wall. This horizontal force is called lateral earth pressure. Care must be taken during construction to avoid having lateral earth pressure from collapsing the foundation wall. Backfilling before the first floor framing and the basement slab are installed typically produces stresses that the wall was not designed to withstand. The following table provides information regarding the lateral earth pressure produced by various types of soils.

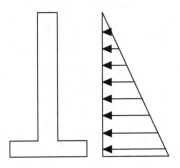

Table 4.11—Soil Loads

Type of Soil in Moist Condition	Soil Load in psf Per Foot of Depth	Soil Load Pounds Per Lineal Foot of 8 ft High Wall
Gravels	35	1120
Gravel–Sand Mixture	35	1120
Clayey Gravels	45	1440
Silty Sands	35	1120
Inorganic Silts	85	2720
Clayey Sands	85	2720

WIND LOAD FAILURE MODES

Three types of failures that can be caused by wind forces are shown. Failure mode 1 depicts a failure that is the result of inadequate foundation connection. A structure must be tied to the foundation by bolts that are embedded in the foundation wall. If there is an inadequate amount of bolts, improperly sized bolts or improper embedment, the wall may lift or slide off the foundation. Failure mode 2 is similar to the previously discussed failure mode except that the roof of the structure is not properly tied to the walls. This problem can be avoided by proper nailing of the roof members to the wall or utilizing tie down straps. Failure mode 3 is the result of the windward side of the roof being suctioned upward. The roof must be properly secured to the walls to prevent this uplift.

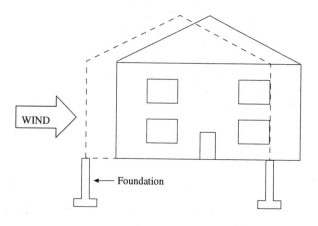

Failure 1

Figure 4.1—Wind Failure

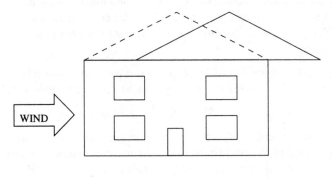

Failure 2

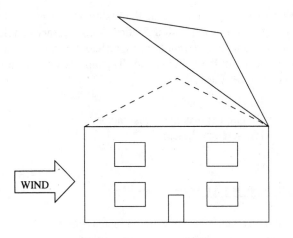

Failure 3

Figure 4.1—(cont.) Wind Failure

The governing building code defines the wind loads that are to be applied to the structure. Once these loads are determined, a structural analysis can be performed to evaluate the stresses in the structure.

The wind pressure that will be experienced by the structure is determined by utilizing the following mathematical expression:

Wind force = $0.00256 \times$ velocity$^2 \times$ modification factors

The first two terms in this expression represent an equation commonly used in physics that the mass of a material times its velocity squared is equal to the force. The building code prescribes the appropriate velocity to place into the equation. The velocity chosen is basically a wind speed that will occur only once every 25, 50, 75 or 100 years.

The force that is determined using the first two terms of the equation is not adequate for design. Modifications are necessary to adjust the forces to account for different terrains, different plan dimensions of buildings, height of the building, etc. The modification factors are from ASCE 7-1993 are

I = Importance factor
C_e = Height, exposure and gust factor
C_p = Pressure coefficient

IMPORTANCE FACTOR

The importance factor provides for an increase in the wind force for essential structures. Essential structures include fire stations, police stations, hospitals and other similar facilities. The theory behind this increase in design forces is the necessity of these structures remaining operational in the event of a disaster.

In contrast, the importance factor does allow for a reduction in wind forces for structures that would represent a low hazard to human life if they failed. These types of structures include temporary facilities, agricultural facilities, and minor storage buildings.

Table 4.12 shows the importance factors for various categories of structures

Table 4.12—Importance Factors

Category	Description	Factor
I	All buildings not in categories II, III, and IV	1.0
II	Buildings that hold 300 or more people in one area	Increase by 15% (1.15)
III	Essential facilities	Increase by 15% (1.15)
IV	Low hazard buildings	Decrease by 10% (0.90)

HEIGHT, EXPOSURE AND GUST FACTOR

The modification factor, C_e, takes into account that different height structures of different shapes in different terrains will experience different wind forces.

Because the nature of the surroundings have an impact on the magnitude of the force, it is necessary to divide terrains into different categories, called exposures.

Exposure A = Center or large cities

Exposure B = Suburban areas

Exposure C = Flat and open country and grasslands

Exposure D = Flat coastal areas subject to wind blowing over bodies of water

Each of these categories has corresponding modification factors that take into account changes in the wind force from the exposure type and turbulence (i.e., gusts). For Exposure A, the modification factors for structures up to 30 ft in height are given as the following:

C_e for Exposure A

Height Above Ground (ft)	C_e
0 — 15	0.28
20	0.33
25	0.36
30	0.38

As a rough rule of thumb, the values of C_e for residential structures in this height range in Exposures B, C, and D are values of C_e listed in the Table multiplied by 2, 3.5 and 4.5, respectively.

PRESSURE COEFFICIENT

The pressure coefficient accounts for the shape of the structure in determining the wind force. For walls, this modification factor is 1.3 for buildings that are fairly square in plan.

The value of C_p for a roof depends on the height and pitch of the roof. For a roof with a 30-degree angle from the horizontal degree, a value of 0.9 providing uplift would be required.

Sample calculations of the wind force at the top of a 15-ft wall under the listed condition is the following:

Case 1: Wind speed: $V = 70$ mph (103 ft/sec)

Low-hazard structure: $I = 0.90$

Center of large city (Exp. A): $C_e = 0.28$

Nearly square in plan $C_p = 1.3$

Wind pressure $= 0.00256 \times V^2 \times I \times C_e \times C_p$
$= 0.00256 \times (103)^2 \times 0.90 \times 0.28 \times 1.3$
$= 8.9$ psf

Case 2: Wind speed: $V = 90$ mph (132 ft/sec)

Essential facility: $I = 1.15$

Open Country (Exp C): $C_e = 0.28 \times 3.5 = 0.98$

Nearly square in plan $C_p = 1.3$

Wind pressure $= 0.00256 \times V^2 \times I \times C_e \times C_p$
$= 0.00256 \times (132)^2 \times 1.15 \times 0.98 \times 1.3$
$= 65.4$ psf

SEISMIC CONSIDERATIONS

Design of a structure for seismic forces presents difficulties beyond those presented by other load conditions. The level of accurate predictions of the time, location and magnitude of an earthquake is presently beyond the capabilities of present day experts. The impact an earthquake will have on a structure depends on the interaction of many different variables. The force that a structure will experience is a function of the type of material used, the height of the building, the underlying soil, the frequency of the earthquake impulse, etc.

Earthquake forces act uniquely on a structure. Wind forces will push directly against a structure, snow loads rest directly on the roof, but earthquake forces do not directly bear on the structure. Earthquakes cause the ground to move, which in turn causes the foundation to move. The movement of the foundation will cause the building to follow the motion. This vibration of the structure causes stresses.

The design philosophy for seismic loads presents a unique twist on the usual design philosophy for other types of loads. Depending on their locale, structures are continuously subjected to snow, wind and live loads. These structures safely resist these loads because they are designed for a magnitude of load that structure will most likely never be subjected to during its life. As an example, consider that in the northern portion of the state of Illinois, the design ground snow load is 25 psf. This translates into between 18 inches and 24 inches of snow. However, in a given winter it would be unlikely that more than 16 inches of snow would accumulate. Thus, this design is for a magnitude greater than what would be expected.

Seismic design is the reverse of this philosophy. Structures are not designed to resist loads greater than will be produced by seismic activity. Instead, they are only designed to withstand less than half of the possible forces.

Therefore, it is an accepted design principle that if a structure is subjected to a significant earthquake, it will possibly be severely damaged. However, it is designed to survive. It is uneconomical to design a structure for the full force of an earthquake. Instead, special construction details are proposed that may damage the earthquake forces but will not allow collapse. For this reason, expert or experienced persons who design for earthquake forces should be contacted when building in a high risk seismic zone.

FORCES

An adequately designed beam must satisfy both serviceability and strength requirements. Serviceability requirements are restrictions that ensure that the beam will perform in a manner that will not be annoying to its users nor cause damage to other components of the structure. Serviceability requirements are for the most part satisfied by limiting the amount of deflection of the beam. Other examples of serviceability requirements include setting minimum dimensions for the beams or providing minimal acceptable limits of material properties.

Strength requirements of members are not related to performance characteristics but are related to the stress in the material. That is, the beam must be sufficiently designed such that it will not fail.

To avoid failure or distress, the beam must be able to support the applied loads. These loads will produce forces at the beam supports, shear forces and bending moments.

The loads imposed on the beam are transferred to the walls or columns that support it. The loads that the beam transfer to the columns or walls are called *support forces*. A beam supported by underdesigned columns or walls is of little value. To design the column or wall, the support forces must be known.

The applied loads will induce a shear force in the beam. A shear force will cause one section of the beam to slip past another section of the beam. This is shown in Figure 4.2.

A beam will deflect downward or sometimes upward when a load is applied. This deflection will increase or decrease along the length of the member, giving it a curved shape. This curved shape, or rotation, will produce stresses that are both compressive and tensile. This is because one side of the beam is experiencing shortening (compressive) while the other side is elongating (tensile). The rotational force causing this stress is called the *bending moment*.

Thus the support reaction is needed to design the supporting column. The shear force and bending moment are needed to design the beam. This chapter provides the maximum values for these three items for a variety of loading and beam conditions. At the end of this chapter illustrative examples are provided.

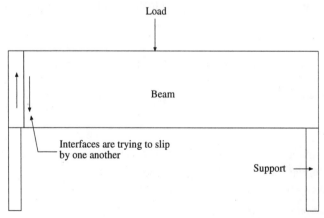

Figure 4.2—Shearing forces in a beam

CASE 1

Description: Uniformly distributed load on entire length of simply supported beam.

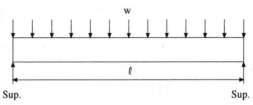

Maximum bending moment = $M = w\ell^2/8$ (Center)

Maximum shear = $V = w\ell/2$ end

Force on left support = $R_L = w\ell/2$

Force on right support = $R_R = w\ell/2$

CASE 2

Description: Uniformly distributed load on half of simply supported beam.

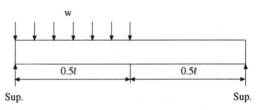

Maximum bending moment = $M = 9w\ell^2/128$ (Near center)

Maximum shear = $V = 3w\ell/8$ (Left end)

Force on left support = $R_L = 3w\ell/8$

Force on right support = $R_R = w\ell/8$

CASE 3

Description: Uniformly distributed load on one-quarter of simply supported beam.

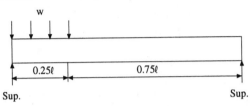

Maximum bending moment = $M = 49w\ell^2/2048$ (Near quarter point)

Maximum shear = $V = 7w\ell/32$ (Left end)

Force on left support = $R_L = 7w\ell/32$

Force on right support = $R_R = w\ell/32$

CASE 4

Description: Concentrated load at midspan of simply supported beam.

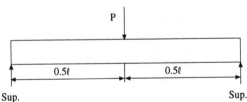

Maximum bending moment = $M = P\ell/4$ (Center)

Maximum shear = $V = P/2$ (Ends)

Force on left support = $R_L = P/2$

Force on right support = $R_R = P/2$

CASE 5

Description: Concentrated load located at 0.4ℓ from end of simply supported beam.

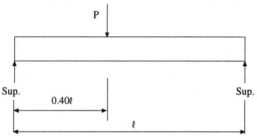

Maximum bending moment = $M = 6P\ell/25$ (At load)

Maximum shear = $V = 3P/5$ (Left side of beam)

Force on left support = $R_L = 3P/5$

Force on right support = $R_R = 2P/5$

CASE 6

Description: Concentrated load located at 0.3ℓ from end of simply supported beam.

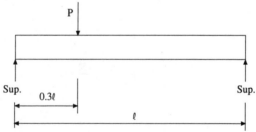

Maximum bending moment = $M = 21P\ell/100$ (At load)

Maximum shear = $V = 7P/10$ (Left end)

Force on left support = $7P/10$

Force on right support = $3P/10$

CASE 7

Description: Concentrated load located at 0.2ℓ from end of simply supported beam.

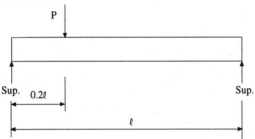

Maximum bending moment = $M = 4P\ell/25$

Maximum shear = $V = 4P/5$ (Left end)

Force on left support = $4P/5$

Force on right end = $P/5$

CASE 8

Description: Uniformly distributed load across entire cantilever beam.

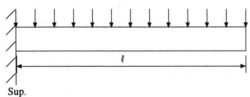

Maximum bending moment = $w\ell^2/2$ (At support)

Maximum shear = $V = w\ell$ (At support)

Force on support = $R = w\ell$

CASE 9

Description: Uniformly distributed load on left half of a cantilever beam.

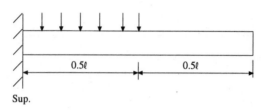

Maximum bending moment = $M = w\ell^2/8$ (At support)

Maximum shear = $V = w\ell/2$ (At support)

Force on support = $R = w\ell/2$

CASE 10

Description: Uniformly distributed load on left one-quarter of cantilever beam.

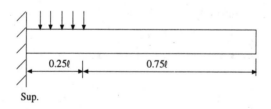

Maximum bending moment = $M = w\ell^2/32$ (At support)

Maximum shear = $V = w\ell/4$ (At support)

Force on support = $R = w\ell/4$

CASE 11

Description: Concentrated load on the end of a cantilever beam.

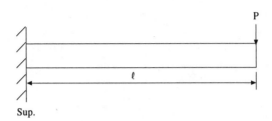

Maximum bending moment = $M = P\ell$ (At support)

Maximum shear = $V = P$ (At support)

Force on support = $R = P$

CASE 12

Description: Concentrated load at the quarter-point from the right end.

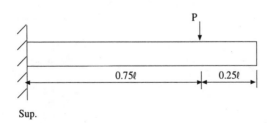

Maximum bending moment = $M = 3P\ell/4$ (At support)

Maximum shear = $V = P$ (At support)

Force at support = $R = P$

CASE 13

Description: Concentrated load at the midpoint of a cantilever beam.

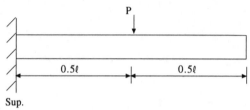

Maximum bending moment = $M = P\ell/2$ (At support)

Maximum shear = $V = P$ (At support)

Force at support = P

CASE 14

Description: Concentrated load at the quarter point from the left end.

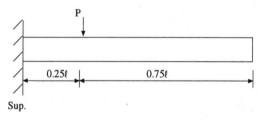

Maximum bending moment = $M = P\ell/4$ (At support)

Maximum shear = $V = P$ (At support)

Force at support = P

CASE 15

Description: Beam with concentrated load on overhang that is 20% of beam length.

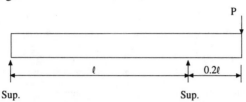

Maximum bending moment = $M = P\ell/5$ (At right support)

Maximum shear = $V = P$ (In overhang)

Force on left support = $R_L = P/5$ (Up)

Force on right support = $R_R = 6P/5$ (Down)

CASE 16

Description: Beam with concentrated load on overhang that is 30% of beam length.

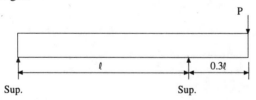

Maximum bending moment = $M = 3P\ell/10$ (At right support)

Maximum shear = $V = P$ (In overhang)

Force on left support = $R_L = 3P/10$ (Up)

Force on right support = $R_R = 13P/10$ (Down)

CASE 17

Description: Beam with concentrated load on overhang that is 40% of beam length.

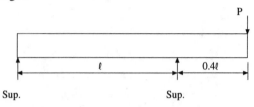

Sup.   Sup.

Maximum bending moment = $M = 2P\ell/5$ (At right support)

Maximum shear = $V = P$ (In overhang)

Force on left support = $2P/5$ (Up)

Force on right support = $7P/5$ (Down)

CASE 18

Description: Uniformly distributed load on overhang of beam. Overhang length is 30% of beam length.

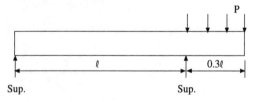

Sup.   Sup.

Maximum bending moment = $M = 9w\ell^2/200$ (At right support)

Maximum shear = $V = 3w\ell/10$ (At right support)

Force on left support = $R_L = 9w\ell/200$ (Up)

Force on right support = $R_R = 69w\ell/200$ (Down)

CASE 19

Description: Uniform load over entire length of beam with overhang. Overhang length is 30% of beam length.

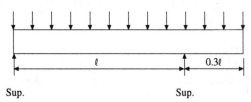

Sup.   Sup.

Maximum bending moment = $M = w\ell^2/10$ (Near midspan)

Maximum shear = $V = 109w\ell/200$ (At right support)

Force at left support = $R_L = 91w\ell/200$

Force at right support = $R_R = 169w\ell/200$

CASE 20

Description: Uniformly distributed load in span for beam with overhang. Overhang length is 30% of beam length.

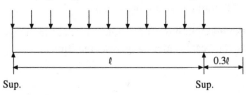

Sup.   Sup.

Maximum bending moment = $M = w\ell^2/8$ (Center)

Maximum shear = $V = w\ell/2$ (At support)

Force on left support = $R_L = w\ell/2$

Force on right support = $R_R = w\ell/2$

CASE 21

Description: Triangular load on simply supported beam.

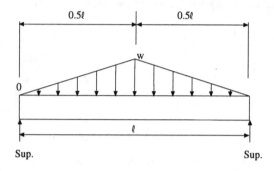

Maximum bending moment = $M = w\ell^2/12$ (At center)

Maximum shear = $V = w\ell/4$ (At support)

Force on each support = $R = w\ell/4$

EXAMPLE 1

This example will compare the results of a load distributed over the entire span to that of a concentrated load.

Determine the maximum bending moment, shear force and support for the following beam under the given conditions.

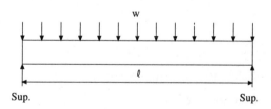

w = Distributed load on beam
= 60 lb/linear ft
= 5 lb/linear in.

ℓ = Length of beam
= 12 ft
= 144 in.

Referring to the information provided for Case 1:

Maximum bending moment = M = $w\ell^2/8$
= $(5)(144)^2/8$
= 12960 in.-lb

Maximum shear = V = $w\ell/2$
= $(5)(144)/2$
= 360 lb

Maximum support force = R = $w\ell/2$
= 360 lb

Compare these values to the results if the distributed load is lumped into one concentrated load at midspan. The concentrated load is equal to the distributed load multiplied by the beam length:

$P = w\ell = (5)(144) = 720$ lb

Referring to the information provided for Case 4, the maximum results are:

Maximum bending moment = M = $P\ell/4$
= $(720)(144)/4$
= 25920 in.-lb

Maximum shear = V = $P/2$
= $720/2$
= 360 lb

$$\text{Maximum support force} = P/2$$
$$= 360 \text{ lb}$$

Comparison of the results show that the shear and support force remain unchanged but the heavy load at midspan doubled the bending moment.

EXAMPLE 2

This example compares the results of a load distributed over the entire span to that of a concentrated load.

Determine the maximum bending moment, shear force and support force for the following cantilever beam under the given conditions.

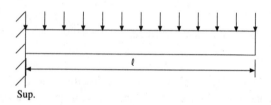

Sup.

w = Distributed load
    = 60 lb/linear ft
    = 5 lb/linear in.

ℓ = Length of beam
    = 4 ft
    = 48 in.

Referring to the information provided for Case 8:

$$\text{Maximum bending moment} = M = w\ell^2/2$$
$$= (5)(48)^2/2$$
$$= 5760 \text{ in.-lb}$$

$$\text{Maximum shear} = V \quad = w\ell$$
$$= (5)(48)$$
$$= 240 \text{ lb}$$

$$\text{Maximum support force} = R \quad = w\ell$$
$$= 240 \text{ lb}$$

Compare these values to the results if the distributed load is lumped into one concentrated load at midspan. The concentrated load is equal to the distributed load multiplied by the beam length:

$$P = w\ell = (5)(48) = 240 \text{ lb}$$

Referring to the information provided for Case 13, the maximum results are the following:

$$\text{Maximum bending moment} = M \quad = P\ell/2$$
$$= (240)(48)/2$$
$$= 5760 \text{ in.-lb}$$

$$\text{Maximum shear} = V = P = 240 \text{ lb}$$

$$\text{Maximum support force} = R = P = 240 \text{ lb}$$

Comparison of the results shows that they produce identical maximums.

The 21 load cases that have been presented thus far in this chapter all had two supports per beam, except the cantilevers. If a beam has more than two supports, it is called a *continuous beam*, as opposed to the two-support condition, which is called a *simple span*. The results for a simple-span beam are significantly different than the same-length beam with three supports. This section presents results for the following cases:

Case 22: Two-span continuous beam with one span loaded. (Three supports)

Case 23: Two-span continuous beam with both spans loaded. (Three supports)

Case 24: Three-span continuous beam with two exterior spans loaded. (Four supports)

Case 25: Three-span continuous beam with three adjacent spans loaded. (Four supports)

CASE 22

Description: Uniform load on one span of a two-span beam.

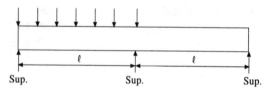

Maximum bending moment = $M = 49w\ell^2/512$ (Near center of loaded span)

Maximum shear = $V = 9w\ell/16$ (At center support)

Force on left support = $R_L = 7w\ell/16$

Force on center support = $R_C = 5w\ell/8$

Force on right support = $R_R = w\ell/16$ (up)

CASE 23

Description: Uniform load on both spans of a two-span beam.

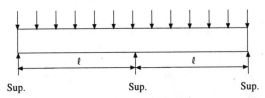

Maximum bending moment = $M = w\ell^2/14$ (Near center of span)

Maximum shear = $V = 5w\ell/8$ (At center support)

Force on end supports = $R_L = R_R = 3w\ell/8$

Force on center support = $R_C = 5w\ell/4$

CASE 24

Description: Uniform load on the end span of a three-span beam.

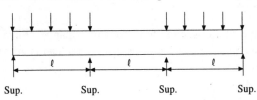

Maximum bending moment = $M = w\ell^2/10$ (Near center of
                                                                                end span)

Maximum shear = $V = 5w\ell/9$ (At center supports)

Force on end supports = $R_E = 5w\ell/11$

Force on interior supports = $R_I = 5w\ell/9$

CASE 25

Description: Uniform load on all spans of a three-span beam.

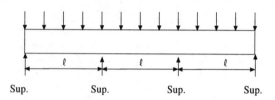

Maximum bending moment = $M = w\ell^2/10$ (Reverse moment   at
                                                    interior support)

Maximum shear = $V = 3w\ell/5$ (At interior support)

Force on end support = $R_e = 2w\ell/5$

Force on interior support = $R_I = 11w\ell/10$

EXAMPLE 3

Compare the interior support force for a two-span and three-span
beam as shown below. Assume the distributed load is the same for
both cases.

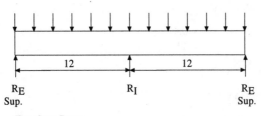

Two-Span Beam

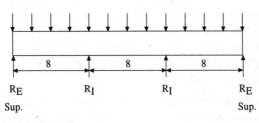

Three-Span Beam

For the two-span beam the interior support force is given by (see Case 23) the following:

$$R_I \quad = 5w\ell/4$$
$$= 5w \, (12 \times 12)/4$$
$$= 180w$$

For the three-span beam the interior support force is given by (see Case 25) the following:

$$R_I \quad = 11w\ell/10$$
$$= 11w \, (8 \times 12)/10$$
$$= 106w$$

Therefore, by the addition of an extra support, the interior support force is reduced by

$$1 - 0.59 = 0.41$$
$$= 41\%$$

DEFLECTIONS

A structural member must have sufficient strength to carry the applied loads with tolerable deflections. Even if a member has sufficient strength it will not be acceptable if there is unsightly deflections in the span. These undesirable deflections will not only look unsightly but they might result in cracking in masonry, plaster and architectural finishes.

Various codes and specifications provide guidelines for the maximum allowable deflections. The following are two of these guidelines.

Floor live load deflection = span length/360

Beams supporting masonry = span length/600

Note that deflection limits are the same regardless of the supporting material used.

Example: A 12-ft long beam is used to support the floor live loads. What is the maximum allowable deflection?

For a 12-ft long beam, the maximum allowable deflection is the span length divided by 360.

$$\text{Maximum deflection} = (12 \times 12)/360$$
$$= 0.4 \text{ in.}$$

The introduction portion of this chapter presents the deflections for 21 different beam conditions. The advanced discussion presents four additional cases.

INTRODUCTION

The following beam conditions are provided in this section:

Case 1: Simply Supported—Uniform Load

Case 2: Simply Supported—Half Beam Loaded

Case 3: Simply Supported—Quarter Beam Loaded

Case 4: Simply Supported—Midspan Load

Case 5: Simply Supported—Load at $0.4L$

Case 6: Simply Supported—Load at $0.3L$

Case 7: Simply Supported—Load at $0.2L$

Case 8: Cantilever—Uniform Load

Case 9: Cantilever—Half Beam Loaded

Case 10: Cantilever—Quarter Beam Loaded

Case 11: Cantilever—End Load

Case 12: Cantilever—Load at $(3/4)L$

Case 13: Cantilever—Load at Midspan

Case 14: Cantilever—Load at $(1/4)L$

Case 15: Overhang (20%)—Load on Overhang

Case 16: Overhang (30%)—Load on Overhang

Case 17: Overhang (40%)—Load on Overhang

Case 18: Overhang (30%)—Distributed Load on Overhang

Case 19: Overhang (30%)—Full Distributed Load

Case 20: Overhang (30%)—Distributed Load in Span

Case 21: Triangular Load

CASE 1

Description: Uniformly distributed load on entire length of simply supported beam.

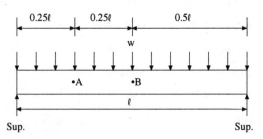

Deflection at A = $w\ell^4/(108EI)$ 71% of maximum

Deflection at B = $w\ell^4/(77EI)$ maximum

EXAMPLE

Calculate the deflection at midspan for the following conditions:

w = Load on beam
        = 55 lb/linear ft
        = 4.6 lb/linear in.

ℓ = Length of beam
        = 12 ft
        = 144 in.

E = Modulus of elasticity
        = 1,400,000 lb/in.^{2}

I = Moment of inertia
        = 178 in.^{4}

Δ_B = $w\ell^4/(108EI)$
        = $(4.6)(144)^4/(77 \times 1,400,000 \times 178)$
        = 0.103 in.
        = Approx. 3/32 in.

CASE 2

Description: Uniformly distributed load on half of simply supported beam.

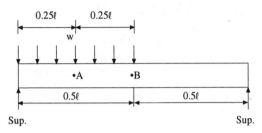

Deflection at A = $w\ell^4/(198EI)$ 77% of maximum

Deflection at B = $w\ell^4/(154EI)$ 99% of maximum

EXAMPLE

Calculate the maximum deflection for the following conditions:

w = Load on beam
      = 55 lb/linear ft
      = 4.6 lb/linear in.

ℓ = Length of beam
      = 12 ft
      = 144 in.

E = Modulus of elasticity
      = 1,400,000 lb/in.

I = Moment of Inertia
      = 178 in.[4]

Δ_B $= w\ell^4/(154EI)$
       $= 4.6(144)^4/(154 \times 1,400,000 \times 178)$
       $= 0.0515$ in.

Δ_B $= 99\%$ of Δ_{max}

Therefore: Δ_{max} $= \Delta_B/0.99$
                          $= 0.052$ in.
                          $=$ Approx. 3/64 in.

CASE 3

Description: Uniformly distributed load on one-quarter of simply supported beam.

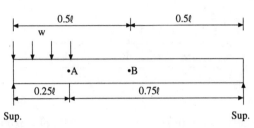

Deflection at A = $w\ell^4/(630EI)$ 83% of maximum

Deflection at B = $w\ell^4(534EI)$ 98% of maximum

EXAMPLE

Calculate the deflection at midspan for the following conditions:

w = Load on beam
        = 55 lb/linear ft
        = 4.6 lb/linear in.

ℓ = Length of beam
        = 12 ft
        = 144 in.

E = Modulus of elasticity
        = 1,400,000 lb/in.

I = Moment of inertia
        = 178 in.^{4}

Δ_B = $w\ell^4/(535EI)$
        = $(4.6)(144)^4/(535 \times 1,400,000 \times 178)$
        = 0.015 in.
        = Approx 1/64 in.

CASE 4

Description: Concentrated load at midspan of simply supported beam.

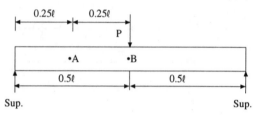

Deflection at A = $P\ell^3/(70EI)$ 69% of maximum

Deflection at B = $P\ell^3/(48EI)$ Maximum

EXAMPLE

Calculate the deflection at midspan for the following conditions:

P	= Concentrated load
	= 10,000 lb

ℓ	= Length of beam
	= 12 ft
	= 144 in.

E	= Modulus of elasticity
	= 29,000,000 lb/in.^{2}

I	= Moment of inertia
	= 48 in.^{4}

Δ_A	= $P\ell/(48EI)$
	= $(10,000)(144)^3/(48 \times 29,000,000 \times 48)$
	= 0.45 in.
	= Approx 1/2 in.

CASE 5

Description: Concentrated load located at 0.4ℓ from end of simply supported beam.

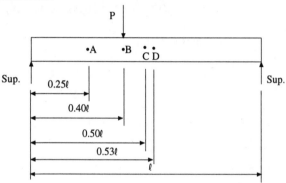

Deflection at A = $P\ell^3/(69EI)$ 73% of maximum

Deflection at B = $P\ell^3/(52EI)$ 97% of maximum

Deflection at C = $P\ell^3/(51EI)$ 99.6% of maximum

Deflection at D = $P\ell^3/(51EI)$ Maximum

EXAMPLE

Calculate the deflection at the quarter point near the left end of the beam for the following conditions:

$$P \quad = \text{Concentrated load}$$
$$= 10,000 \text{ lb}$$

$$\ell \quad = \text{Length of beam}$$
$$= 12 \text{ ft}$$
$$= 144 \text{ in.}$$

$$E \quad = \text{Modulus of elasticity}$$
$$= 29,000,000 \text{ lb/in.}^2$$

I = Moment of inertia
           = 48 in.^{4}

Δ_A = $P\ell^3/(69EI)$
           = $(10,000)(144)^3/(69 \times 29,000,000 \times 48)$
           = 0.31
           = Approx. 5/16 in.

CASE 6

Description: Concentrated load located at 0.3ℓ from end of simply supported beam.

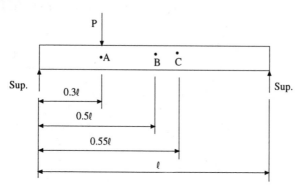

Deflection at A = $P\ell^3/(68EI)$ 88% of maximum

Deflection at B = $P\ell^3/(61EI)$ 99% of maximum

Deflection at C = $P\ell^3/(60EI)$ Maximum

EXAMPLE

Calculate deflection at midspan for following conditions:

P = Concentrated load
              = 10,000 lb

ℓ = Length of beam
              = 12 ft
              = 144 in.

E = Modulus of elasticity
              = 29,000,000 lb/in.^{2}

I = Moment of inertia
         = 48 in.^{4}

Δ_B = $P\ell^3/(61EI)$
         = $(10{,}000)(144)^3/(61 \times 29{,}000{,}000 \times 48)$
         = 0.35 in.
         = Approx. 11/32 in.

CASE 7

Description: Concentrated load located at 0.2ℓ from end of simply supported beam.

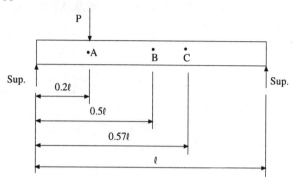

Deflection at A = $P\ell^3/(117EI)$ 71% of maximum

Deflection at B = $P\ell^3/(85EI)$ 98% of maximum

Deflection at C = $P\ell^3/(83EI)$ Maximum

EXAMPLE

Calculate deflection at the location of the concentrated load for the following conditions:

P = Concentrated load
             = 10,000 lb

ℓ = Length of beam
             = 12 ft
             = 144 in.

E = Modulus of elasticity
             = 29,000,000 lb/in.^{2}

I = Moment of inertia
      = 48 in.^{4}

Δ_A = $P\ell^3/(117EI)$
      = $(10,000)(144)^3/(117 \times 29,000,000 \times 48)$
      = 0.18 in.
      = Approx. 3/16 in.

CASE 8

Description: Uniformly distributed load across entire cantilever beam.

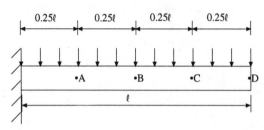

Deflection at A = $w\ell^4/(76EI)$ 11% of maximum

Deflection at B = $w\ell^4/(23EI)$ 35% of maximum

Deflection at C = $w\ell^4/(12EI)$ 67% of maximum

Deflection at D = $w\ell^4/(8EI)$ Maximum

EXAMPLE

Calculate the deflection at the free end of the beam for the following conditions:

w = Distributed load on beam
         = 500 lb/linear ft
         = 41.7 lb/linear in.

ℓ = Length of cantilever beam
         = 4 ft
         = 48 in.

E = Modulus of elasticity
         = 1,400,000 lb/in.^{2}

I = Moment of inertia
          = 178 in.^{4}

Δ_D = $w\ell^4/(8EI)$
          = $(41.7)(48)^4/(8 \times 1,400,000 \times 178)$
          = 0.11 in.
          = Approx. 1/8 in.

CASE 9

Description: Uniformly distributed load on left half of a cantilever beam.

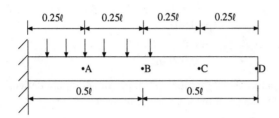

Deflection at A = $w\ell^4/(361EI)$ 15% of maximum

Deflection at B = $w\ell^4/(128EI)$ 43% of maximum

Deflection at C = $w\ell^4/(77EI)$ 71% of maximum

Deflection at D = $w\ell^4/(55EI)$ Maximum

EXAMPLE

Calculate the additional deflection that occurs after the load has ended for the following conditions:

w = Distributed load on beam
       = 500 lb/linear ft
       = 41.7 lb/linear in.

ℓ = Length of cantilever beam
       = 4 ft
       = 48 in.

E = Modulus of elasticity
       = 1,400,000 lb/in.^{2}

I = Moment of inertia
       = 178 in.^{4}

This calculation is performed by determining the difference between the deflection at the end and the deflection at midspan.

$$\Delta_D \quad = w\ell^4/(55EI)$$
$$= (41.7)(48)^4/(55 \times 1,400,000 \times 178)$$
$$= 0.016$$

$$\Delta_B \quad = w\ell^4/(128EI)$$
$$= (41.7)(48)^4/(128 \times 1,400,000 \times 178)$$
$$= 0.007$$

$$\Delta_D - \Delta_B \quad = 0.016 - 0.007$$
$$= 0.009 \text{ in.}$$

Approximately 0.009 (approx. 1/100 in.) of additional deflection occurs after the point the load has terminated.

CASE 10

Description: Uniformly distributed load on left one-quarter of cantilever beam.

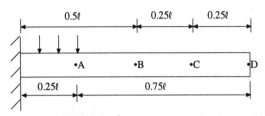

Deflection at A = $w\ell^4/(2048EI)$	20% of maximum
Deflection at B = $w\ell^4/(878EI)$	47% of maximum
Deflection at C = $w\ell^4/(559EI)$	73% of maximum
Deflection at D = $w\ell^4/(410EI)$	Maximum

EXAMPLE

Calculate the deflection at the location where the distributed load ends for the following condition:

w = Distributed load on beam
= 500 lb/linear ft
= 41.7 lb/linear in.

ℓ = Length of cantilever beam
= 4 ft
= 48 in.

E = Modulus of elasticity
= 1,400,000 lb/in.^{2}

I = Moment of inertia
= 178 in.^{4}

Δ_A = $w\ell^4/(2048EI)$
= $(41.7)(48)^4/(2048 \times 1,400,000 \times 178)$
= 0.0004 in.
= Approx. 1/2000 in.

CASE 11

Description: Concentrated load on end of a cantilever beam.

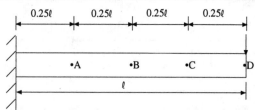

Deflection at A = $P\ell^3/(35EI)$ 9% of maximum

Deflection at B = $P\ell^3/(10EI)$ 31% of maximum

Deflection at C = $P\ell^3/(5EI)$ 63% of maximum

Deflection at D = $P\ell^3/(3EI)$ Maximum

EXAMPLE

Calculate the maximum deflection for the following conditions:

P = Concentrated load
   = 1,000 lb

ℓ = Length of beam
   = 4 ft
   = 48 in.

E = Modulus of elasticity
   = 1,400,000 lb/in.^{2}

I = Moment of inertia
   = 48 in.^{4}

Δ_D = $P\ell^3/(3EI)$
   = $(1000)(48)^3/(3 \times 1,400,000 \times 48)$
   = 0.55 in.
   = Approx. 1/2 in.

CASE 12

Description: Concentrated load at quarter point from the right end.

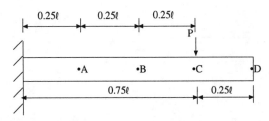

Deflection at A = $P\ell^3/(48EI)$ 10% of maximum

Deflection at B = $P\ell^3/(14EI)$ 35% of maximum

Deflection at C = $P\ell^3/(7EI)$ 67% of maximum

Deflection at D = $P\ell^3/(5EI)$ Maximum

EXAMPLE

Calculate the maximum deflection for the following conditions:

p = Concentrated load
      = 1000 lb

ℓ = Length of beam
      = 4 ft
      = 48 in.

E = Modulus of elasticity
      = 1,400,000 lb/in.^{2}

I = Moment of inertia
      = 48 in.^{4}

Δ_D = $P\ell^3/(5EI)$
      = $(1,000)(48)^3/(5 \times 1,400,000 \times 48)$
      = 0.33 in.
      = Approx. 5/16 in.

CASE 13

Description: Concentrated load at midpoint of a cantilever beam.

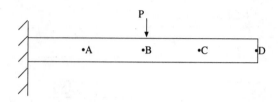

Deflection at A = $P\ell^3/(77EI)$ 13% of maximum

Deflection at B = $P\ell^3/(24EI)$ 42% of maximum

Deflection at C = $P\ell^3/(14EI)$ 71% of maximum

Deflection at D = $P\ell^3/(10EI)$ Maximum

EXAMPLE

Calculate the load at midspan needed to cause a 1-in. deflection at the free end for the following conditions.

ℓ = Length of beam
  = 6 ft
  = 72 in.

E = Modulus of elasticity
  = 1,400,000 lb/in.^{2}

I = Moment of inertia
  = 48 in.^{4}

The deflection at the end of the beam is given by the following equation:

$$\Delta_D = P\ell^3/(10EI)$$

Including the information that is known into the equation gives

$$1 = P(72)^3/(10 \times 1{,}400{,}000 \times 48)$$

$$1 = \frac{373248P}{672{,}000{,}000}$$

$$1 = 0.000555P$$

$$P = 1/0.000555$$

$$P = 1800 \text{ lb}$$

CASE 14

Description: Concentrated load at quarter point from the left end.

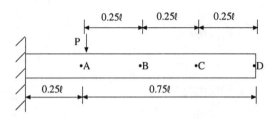

Deflection at A = $P\ell^3/(192EI)$ 18% of maximum

Deflection at B = $P\ell^3/(77EI)$ 45% of maximum

Deflection at C = $P\ell^3/(48EI)$ 73% of maximum

Deflection at D = $P\ell^3/(35EI)$ Maximum

EXAMPLE

Calculate the maximum load that can be applied at Point A to cause a 1-in. deflection at the free end for the following conditions:

$$\ell \qquad = \text{Length of beam}$$
$$= 6 \text{ ft}$$
$$= 72 \text{ in.}$$

$$E \qquad = \text{Modulus of elasticity}$$
$$= 1,400,000 \text{ lb/in.}^2$$

$$I \qquad = \text{Moment of inertia}$$
$$= 48 \text{ in.}^4$$

The deflection at the end of the beam is given by the following equation:

$$\Delta_D = P\ell^3/(35EI)$$

Substituting the information that is known into the equation gives

$$1 \text{ in.} = P(72)^3/(35 \times 1,400,000 \times 48)$$

$$1 \text{ in.} = \frac{373248P}{2,352,000,000,000}$$

$$1 \text{ in.} = 0.00016P$$

$$P = 6301 \text{ lb}$$

CASE 15

Description: Beam with concentrated load on overhang that is 20% of beam length.

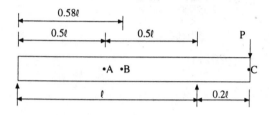

Upward deflection at A = $P\ell^3/(80EI)$

Upward deflection at B(0.58ℓ) = $P\ell^3/(78EI)$ Maximum in span

Downward deflection at C = $P\ell^3/(63EI)$ Maximum in overhang

EXAMPLE

Calculate the distance that the center of the beam moves upward under the following conditions:

$$P \quad = \text{Concentrated load}$$
$$= 10,000 \text{ lb}$$

$$\ell \quad = \text{Length of beam}$$
$$= 10 \text{ ft}$$
$$= 120 \text{ in.}$$

$$\text{Overhang} \quad = 2 \text{ ft}$$
$$= 24 \text{ in.}$$

$$E \quad = \text{Modulus of elasticity}$$
$$= 1,400,000 \text{ lb/in.}^2$$

I = Moment of inertia
        = 178 in.^{4}

Δ_A = $P\ell^3/(80EI)$
        = $(10,000)(120)^3/(80 \times 1,400,000 \times 178)$
        = 0.87 in.
        = 7/8 in.

CASE 16

Description: Beam with concentrated load on overhang that is 30% of beam length.

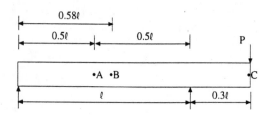

Upward deflection at A = $P\ell^3/(53EI)$

Upward deflection at B$(0.58\ell) = P\ell^3/(52EI)$ Maximum in span

Downward deflection at C = $P\ell^3/(26EI)$ Maximum in overhang

EXAMPLE

Calculate the maximum downward deflection in the beam under the following condition:

$$P \qquad = \text{Concentrated load}$$
$$= 10,000 \text{ lb}$$

$$\ell \qquad = \text{Length of beam}$$
$$= 10 \text{ ft}$$
$$= 120 \text{ in.}$$

$$\text{Overhang} \qquad = 3 \text{ ft}$$
$$= 36 \text{ in.}$$

$$E \qquad = \text{Modulus of elasticity}$$
$$= 1,400,000 \text{ lb/in.}^2$$

I = Moment of inertia
        = 178 in.^{4}

Δ_C = $P\ell^3/(26EI)$
        = $(10,000)(120)^3/(26 \times 1,400,000 \times 178)$
        = 2.67 in.
        = 2-5/8 in.

CASE 17

Description: Beam with concentrated load on overhang that is 40% of beam length.

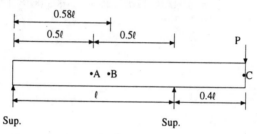

Upward deflection at A = $P\ell^3/(40EI)$

Upward deflection at B(0.58ℓ) = $P\ell^3/(39EI)$ Maximum in span

Downward deflection at C = $P\ell^3/(13EI)$ Maximum in overhang

EXAMPLE

Calculate the maximum amount the beam will rise under the following conditions:

P = Concentrated load
        = 10,000 lb

ℓ = Length of beam
        = 10 ft
        = 120 in.

Overhang   = 4 ft
             = 48 in.

E = Modulus of elasticity
        = 1,400,000 lb/in.^{2}

I = Moment of inertia
= 178 in.^{4}

Δ_C = $P\ell^3/(39EI)$
= $(10,000)(120)^3/(39 \times 1,400,000 \times 178)$
= 1.78 in.
= 1-3/4 in.

CASE 18

Description: Uniformly distributed load on overhang of beam.
Overhang length is 30% of beam length.

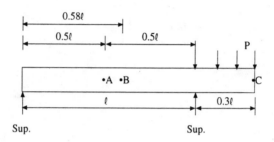

Upward deflection A = $w\ell^4/(356EI)$

Upward deflection B(0.58ℓ) = $w\ell^4/(346EI)$ Maximum in span

Downward deflection C = $w\ell^4/(181EI)$ Maximum

EXAMPLE

Calculate the maximum downward deflection in the beam under the
following conditions:

$$
\begin{aligned}
w \quad &= \text{Uniformly distributed load} \\
&= 3340 \text{ lb/linear ft} \\
&= 278 \text{ lb/linear in.}
\end{aligned}
$$

$$
\begin{aligned}
\ell \quad &= \text{Length of beam} \\
&= 10 \text{ ft} \\
&= 120 \text{ in.}
\end{aligned}
$$

$$
\begin{aligned}
\text{Overhang} \quad &= 3 \text{ ft} \\
&= 36 \text{ in.}
\end{aligned}
$$

E = Modulus of elasticity
        = 1,400,000 lb/in.^{2}

I = Moment of inertia
        = 178 in.^{4}

Δ_B = $w\ell^4/(181EI)$
        = $(278)(120)^4/(181 \times 1{,}400{,}000 \times 178)$
        = 1.28 in.
        = 1-1/4 in.

CASE 19

Description: Uniformly load over entire length of beam with overhang. Overhang length is 30% of beam length.

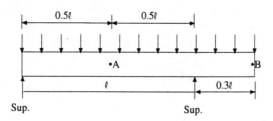

Downward deflection at A = $w\ell^4/(98EI)$

Upward deflection at B = $w\ell^4/(143EI)$

EXAMPLE

Calculate the maximum upward deflection at B under the following conditions:

w = Uniformly distributed load
     = 3340 lb/linear ft
     = 278 lb/linear in.

ℓ = Length of beam
     = 10 ft
     = 120 in.

Overhang   = 3 ft
          = 36 in.

E = Modulus of elasticity
     = 1,400,000 lb/in.^{2}

I = Moment of inertia
     = 178 in.^{4}

Δ_B = $w\ell^4/(143EI)$
     = $(278)(120)^4/(143 \times 1,400,000 \times 178)$
     = 1.62 in.
     = 1-5/8 in.

CASE 20

Description: Uniformly distributed load in span for beam with overhang. Overhang length is 30% of beam length.

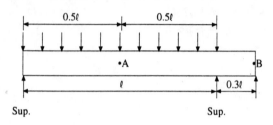

Downward deflection at A = $w\ell^4/(77EI)$

Upward deflection at B = $w\ell^4/(80EI)$

EXAMPLE

Calculate the maximum upward deflection at B under the following conditions:

w = Uniformly distributed load
          = 3340 lb/linear ft
          = 278 lb/linear in.

ℓ = Length of beam
          = 10 ft
          = 120 in.

Overhang   = 3 ft
          = 36 in.

E = Modulus of elasticity
          = 1,400,000 lb/in.^{2}

I = Moment of inertia
            = 178 in.^{4}

Δ_B = $w\ell^4/(80EI)$
            = $(278)(120)^4/(80 \times 1,400,000 \times 178)$
            = 2.89 in.
            = 2-7/8 in.

CASE 21

Description: Triangular load on simply supported beams.

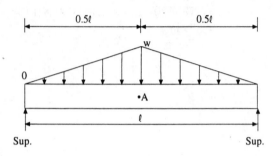

Deflection at A = $Wl^3/(60EI)$

EXAMPLE

Calculate the deflection at midspan for the following conditions:

W = Total load on beam
= $wl/2$

w = 660 lb/linear ft
= 55 lb/linear in.
= (55)(144)/2
= 3960 lb

l = Length of beam
= 12 ft
= 144 in.

E = Modulus of elasticity
= 1,400,000 lb/in.^{2}

I = Moment of inertia
            = 178 in.^{4}

Δ_A = $(3960)(144)^3/(60 \times 1,400,000 \times 178)$
            = 0.79 in.
            = Approx. 3/4 in.

The 21 load cases that have been presented thus far in this chapter all had two supports per beam, except for the cantilever beam. If a beam has more than two supports, it is called a *continuous beam,* as opposed to the two-support condition, which is called a *simple span.* The results for a simple span beam is significantly different than the same-length beam with three supports. This section will present results for the following cases:

   Case 22: Two-span continuous beam with one span loaded. (Three supports)

   Case 23: Two-span continuous beam with both spans loaded. (Three supports)

   Case 24: Three-span continuous beam with two exterior spans loaded. (Four supports)

   Case 25: Three-span continuous beam with three spans loaded. (Four supports)

CASE 22

Description: Two-span continuous beam with uniform load in one span.

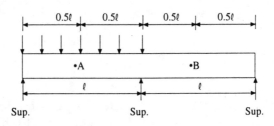

Downward deflection at A = $w\ell^4/(110EI)$

Upward deflection at B = $w\ell^4/(250EI)$

EXAMPLE

The loading has caused the deflection at point A to be 1 in. What is the upward deflection at B, under the following conditions?

$$\ell \quad = \text{Span length}$$
$$= 6 \text{ ft}$$
$$= 72 \text{ in.}$$

$$E \quad = \text{Modulus of elasticity}$$
$$= 1{,}400{,}000 \text{ lb/in.}^2$$

$$I \quad = \text{Moment of inertia}$$
$$= 178 \text{ in.}^4$$

$$\Delta_A = \quad 1 \text{ in.} = w\ell^4/(110EI)$$
$$1 \text{ in.} = (w)(72)^4/(110 \times 1{,}400{,}000 \times 178)$$
$$1 \text{ in.} = w/1020$$
$$w = 1020 \text{ lb/linear in.}$$

The upward deflection at Point B with a load of 1020 lb/linear in. is

$$\Delta_B = w\ell^4/(250EI)$$
$$= (1020)(72)^4/(250 \times 1,400,000 \times 178)$$
$$= 0.44 \text{ in.}$$
$$= \text{Approx. } 7/16 \text{ in. upward}$$

CASE 23

Description: Two-span continuous beam with uniform load in both spans.

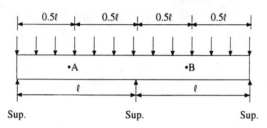

Deflection at A = $w\ell^4/(192EI)$

Deflection at B = Deflection at A

EXAMPLE

Determine the uniform load needed to cause a 1-in. deflection at point A under the following conditions:

$$\ell \qquad = \text{Span length}$$
$$= 6 \text{ ft}$$
$$= 72 \text{ in.}$$

$$E \qquad = \text{Modulus of elasticity}$$
$$= 1,400,000 \text{ lb/in.}^2$$

$$I \qquad = \text{Moment of inertia}$$
$$= 178 \text{ in.}^4$$

$$\Delta_A = \quad 1 = w\ell^4/(192EI)$$
$$1 = w(72)^4/(192 \times 1,400,000 \times 178)$$
$$1 = w/1780$$
$$w = 1780 \text{ lb/linear in.}$$
$$= 148 \text{ lb/linear ft}$$

CASE 24

Description: Three-span continuous beam with two spans loaded.

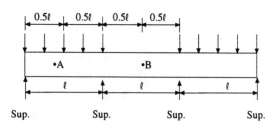

Downward deflection at A $= w\ell^4/(100EI)$

Upward deflection at B $= w\ell^4/(160EI)$

EXAMPLE

What load will result in a 1/2-in. deflection at A under the following conditions:

ℓ = Span length
       = 10 ft
       = 120 in.

E = Modulus of elasticity
       = 1,400,000 lb/in.^{2}

I = Moment of inertia
       = 178 in.^{4}

$\Delta_A =$ $0.5 = w\ell^4/(100EI)$
           $0.5 = w(120)^4/(100 \times 1,400,000 \times 178)$
           $0.5 = w/120$
        w = 60 lb/linear in.
             = 5 lb/linear ft

CASE 25

Description: Three-span continuous beam with all spans loaded.

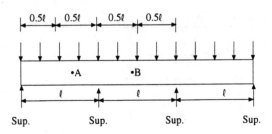

Deflection at A = $w\ell^4/(145EI)$

Deflection at B = $w\ell^4/(1927EI)$

EXAMPLE

If the deflection at A is 1 in., what is the deflection at B?

$$\Delta_A = 1 \text{ in.} = w\ell^4/(145EI)$$
$$145 = w\ell^4/(EI)$$

$$\Delta_B = w\ell^4/(1927EI)$$
$$= 1/1927 \times w\ell^4/EI$$

Substitute 145 for $w\ell^4/(EI)$

$$\Delta_B = 1/1927 \times 145$$

$$\Delta_B = 145/1927$$
$$= 0.08 \text{ in.}$$

Deflections were presented for 21 different load cases. In the advanced discussion of this chapter, more complicated conditions are also presented. Note that all of these loading conditions can be used in combination with one another to determine deflections of many other loading conditions. Combining these load conditions is called *superpositioning*. The theory of superpositioning provides that the deflection caused by several loads is equal to the deflection determined by summing the deflection of each load separately. This theory is explained in the following example.

Determine the deflection at midspan for the beam shown with the given conditions.

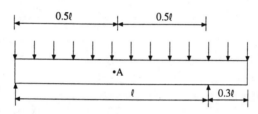

w = Distributed load
       = 1400 lb/linear ft
       = 117 lb/linear in.

ℓ = Length of beam
       = 10 ft
       = 120 in.

Overhang length   = 3 ft
                    = 36 in.

E = Modulus of elasticity
       = 1,400,000 lb/in.^{2}

I = Moment of inertia
       = 178 in.^{4}

With the information provided in Case 19 utilized, the deflection at midspan is

$$\Delta \quad = w\ell^4/(98EI)$$
$$= (117)(120)^4/(98 \times 1,400,000 \times 178)$$
$$= 1.0 \text{ in.}$$

This same result can be obtained by combining the information provided in Cases 18 and 20. The deflection determined using Case 18 is

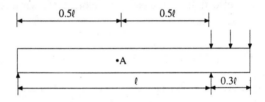

Upward deflection at A $= w\ell^4/(356EI)$
$$= (117)(120)^4/(356 \times 1,400,000 \times 178)$$
$$= 0.3 \text{ in. upward}$$

The deflection determined using Case 20 is

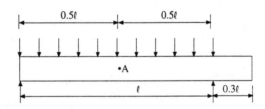

Downward deflection at A $= w\ell^4/(77EI)$
$$= (117)(120)^4/(77 \times 1,400,000 \times 178)$$
$$= 1.3 \text{ in. down}$$

Adding these two deflection [1.3 + (-0.3)] = 1.0 in. The following provides a summary of the previous calculations.

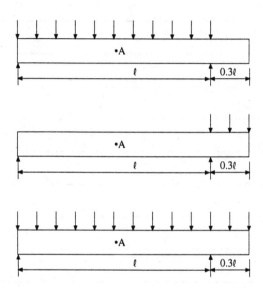

NOTES:

NOTES:

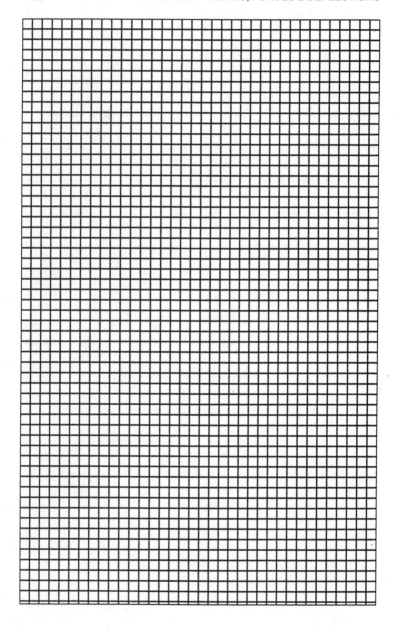

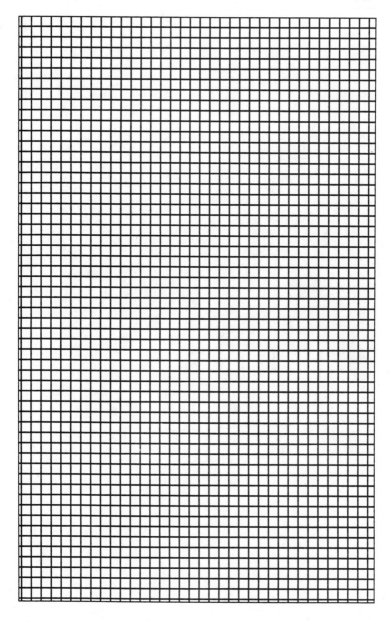

Chapter 5
Concrete

TIPS & INFORMATION

- The more water that is added to the concrete mix, the weaker the concrete will be. Avoid adding water to the truck at the site unless the strength reduction from its inclusion has been considered.

- Concrete is a relatively cheap material. The incremental cost for purchasing a concrete of a slightly higher strength is minimal.

- Steel mesh reinforcing is included in a slab on grade to reduce crack widths. For the mesh to be effective, it must be in the top third of the slab thickness. If the mesh is pushed down to the bottom of the slab during concrete placement, it serves no purpose.

- Avoid overworking the slab surface during finishing operations. Excessive trowling will result in water bleeding to the surface. This water will result in a reduced strength on the top of the slab and spalling will occur.

- Cement and concrete are not the same. Cement when it is mixed with water will bind the stone and sand to form concrete.

- The concrete mixture contains lime. Workers should not have their skin directly exposed to the concrete mix.

- In most residential projects, steel beams will be supported in pockets at the top of the concrete wall. The size of the steel beam should be determined before placing the blockout in the form work to avoid having to make the pocket deeper at a later date.

- A concrete truck can transport as much as 10 cubic yards of concrete, although typically, no more than eight cubic yards is transported.

- Avoid placement of concrete in hot temperatures, particularly when it is windy. These conditions cause the moisture to dissipate from the concrete faster and may allow cracking.

TERMINOLOGY

Accelerator	A substance that is added to the concrete mix that results in an increase in the rate of time it takes for the concrete to harden.
Admixture	Any material besides aggregate, cement and water that is added to the concrete mix to change a property of the mixture.
Aggregate	Sand or gravel that is mixed with cement and water to make concrete.
Air Content	Air voids in concrete which makes concrete more durable by allowing for the expansion of water in localized areas without causing surface damage and improves workability.
Anchor Bolt	A bolt that is drilled or cast into concrete. The bolt is later used to attach other structural elements to the concrete.
Bag of Cement	A measure of cement equal to 94 pounds.
Batch	The amount of concrete mixed at one time.
Blockout	A space in a concrete member that is being constructed to prevent concrete from entering that space.
Bug Holes	The small voids in the surface of concrete that are the result of air bubbles that formed during placement.
Bush Hammer	Jack hammering that produces a rough surface on a concrete surface.
Buttering	Spreading of mortar on the surface of masonry.

Central Mix   Concrete that is mixed in a stationary mixer and
                   then transported to the site.

Cold Joint   A joint in concrete that results from discon-
                   tinuing the placement of concrete and then
                   continuing it after the original pour has hardened.

Compaction   The reduction of volume in concrete by using
                   vibrators for the purpose of eliminating voids.

Concrete   A material that consists of cement, water and
                   aggregate that is placed in a liquid state and then
                   hardens into a solid state.

Crazing   Shallow closely spaced cracks at irregular
                   intervals on concrete surfaces.

Curing   Keeping concrete at a favorable temperature and
                   moisture to allow for hydration of the cement.

Curing   A material applied to the surface of concrete to
Compound   reduce the evaporation of water and allow for
                   better curing.

Curling   The warping of the edge of a concrete panel of a
                   slab on grade. This condition occurs because the
                   top of the slab dries faster than the portion placed
                   on the ground.

Dampproofing   A substance that is used to coat the surface of
                   concrete to prevent the passage of moisture
                   through the material.

Dowel   A steel rod that extends into two adjoining pieces
                   of concrete for the purpose of preventing relative
                   movement of the members.

Dry Mix   Prepackaged concrete that only needs the addition
                   of water.

Efflorescence	A chloride stain that resulted from water that exited the concrete and dried on the exterior surface causing staining.
Falsework	A temporary structure that supports the concrete work being performed.
Fin	A small, thin piece of concrete that protrudes from the surface where the concrete forms abutted each other.
Form	A structure in which concrete is poured into for the purpose of supporting the concrete in the desired shape until it hardens.
Form Grease	Oil that is placed onto the inside of concrete formwork to prevent the formwork from sticking to the concrete.
Green Concrete	Concrete that has set but has not completely hardened.
Keyway	A groove placed in concrete before it fully hardens for the purpose of providing shear strength between the two pours.
Mud Slab	A thin layer of concrete poured on soil.
Nailer	A wood member buried in concrete to allow for attachment by nailing.
Portland Cement	A naturally occurring limestone that is heated and then ground to a fine powder that is used to provide the binding material in concrete.
Steel Reinforcing	Steel rods in concrete for the purpose of assisting the concrete in carrying the tensile forces.
Scaling	Flaking of the surface of concrete.

Slabjacking	The upward movement of a concrete slab by lifting the slab or raising it by pressure injecting.
Spall	A fragment of concrete broken away from the surface.
Tieback	A rod fastened to a rigid weight to prevent the item attached to the other end of the rod from moving.
Tilt-up	The building of a structure by forming members on a horizontal surface plane and then lifting them into their vertical position.
Vapor Barrier	Usually sheets of plastic placed beneath a slab on grade to prevent the upward travel of moisture into the usable space.
Vibration	Agitating concrete for the purpose of consolidating wet concrete in the forms.
Water Reducing Agent	A product added to concrete that increases its workability without having to add more water.
Workability	A property of concrete that defines the ease of which the concrete can be placed.

TYPICAL CODE REQUIREMENTS

CONCRETE PLACEMENT

- Concrete strength shall be based on a 28 day strength unless otherwise specified.

- Strength test samples shall be taken at least once a day and for each 150 cubic yards of concrete.

- The strength test shall be the average of two cylinders made from the same sample.

- Strength requirements shall not be satisfied if a sample is 500 psi below that specified.

- Strength requirements shall not be satisfied if the average of three tests are below that specified.

- Concrete shall be deposited as close as possible to its final position.

- Reinforcement and forms shall be cleaned of ice, coatings and debris prior to concrete placement.

- Concrete that has reached initial set shall not be retempered.

- Concrete shall be thoroughly consolidated around reinforcing bars and into corners of forms.

- Concrete shall be maintained above 50 degrees for at least the first seven days after placement.

- Concrete shall be maintained in a moist condition for at least seven days after placement.

- Slabs shall have control joints with a depth of 1/4 of the slab thickness and not more than 30 feet apart.

- Control joints may not be needed if code specified welded wire mesh is used in the slab.

- A 4-inch base of graded gravel or crushed stone shall be placed beneath the slab on grade.

SLABS

- Floor slabs shall be at least 3 ½ inches.

- Slab on grades shall have a 6 mil polyethylene vapor retarder beneath.

- Vapor retarders are typically not needed beneath slabs not enclosed in living space.

MATERIALS

- Shall comply with American Concrete Institute Building Code ACI 318.

- Cement shall meet ASTM C150.

- Aggregate shall meet ASTM C33.

- Aggregate shall not be larger than 20% of the width between forms.

- Aggregate shall not be larger than 33% of the slab thickness.

- Aggregate shall not exceed 75% of the distance between reinforcing bars.

FORMING CONCRETE

- Forms shall be sufficiently leak free.

- Forms shall be braced to prevent failure and maintain dimensions.

- Pipes embedded in structural members shall not be larger than 1/3 the overall thickness of the concrete member.

REINFORCEMENT

- All reinforcement shall be bent cold.

- Reinforcement placed in forms shall be secured against displacement.

- Concrete cover shall be as follows:

Cast directly against ground	3 inches
Concrete exposed to weather	2 inches
Smaller bars in concrete exposed to weather (#5 or smaller)	1 ½ inches
Concrete not exposed to weather	3/4 inches

DETERMINING QUANTITIES FOR A CONCRETE MIX

The following presents the determination of the quantities for a concrete mix. It is presented for illustration purposes only and should not be relied upon for actual use.

BASIC ASSUMPTIONS

Specified strength: 3500 psi
Mix design is based on approximately one cubic yard
Air entrained concrete is to be used
Dry weight of aggregate is estimated to be 100 pounds
per cubic foot

1. Determine Target Strength

The design mix should produce a strength that is 1200 psi greater than the specified strength. Target strength is 3500 psi + 1200 psi = 4700 psi.

2. Identify Desirable Slump and Maximum Nominal Aggregate Size

A typical slump for residential concrete is 3 inches. Assume for this example the maximum nominal aggregate size is 1 inch.

3. Estimate Required Water

A very rough rule of thumb is that 3/4 inch nominal maximum size aggregate requires 305 pounds of water. For each 1/4 inch increase in nominal maximum aggregate size this amount should be reduced by 10 pounds. The amount of water needed for this mix is

305 pounds - 10 pounds = 295 pounds

4. Estimate Required Cement

A very rough rule of thumb for air entrained concrete is to use a weight of water multiplied by an adjustment factor. The multiplying factor is given by

Multiplying Factor = 1/(0.89 - Concrete Strength/10,000)

For 4700 psi concrete
this factor   = 1/(0.89 - 4700/10,000)
                    = 1/(0.89 - 0.47)
                    = 1/0.42
                    = 2.38

Therefore a weight of cement of 295 pounds × 2.38 = 702 pounds is needed.

5. Estimate Amount of Course Aggregate Required

The weight of aggregate needed for each cubic yard of concrete is 1850 pounds for 1 inch nominal maximum aggregate. This should be increased (or correspondingly decreased) by 80 pounds for every 1/4 inch increase in maximum aggregate size.

Therefore use 1850 pounds of aggregate.

6. Amount of Sand Required

The total weight of a cubic yard of a concrete mix is typically around 4000 pounds. Subtracting out the weights of the items previously determined will give the weight of sand needed.

Water	295 pounds
Cement	702 pounds
Aggregate	1850 pounds
Total	2847 pounds

Required weight of sand needed = 4000 - 2847 = 1153 pounds.

CURING OF CONCRETE

GENERAL INFORMATION

- Proper curing is needed so that enough moisture remains in the concrete such that it can fully develop its strength.

- Proper curing is achieved by reducing the rate of evaporation from the surface of the concrete.

- Evaporation of water from the surface of the concrete can be reduced by watering the surface, covering the surface with plastic or by sealing in the moisture by the use of chemicals.

- The following increases the evaporation rate of moisture and should be avoided or controlled: Higher air temperatures, higher wind velocities, higher concrete temperatures and lower relative humidity.

Table 5.1—Methods of Curing

Method	Procedure	Advantages	Disadvantages
Burlap	Lay wet burlap on concrete surface.	Cheap and easy to apply.	Stays wet only a limited time.
Water Spray	Use lawn sprinkler to continually wet concrete surface.	Provides excellent protection against evaporation.	High water costs and potential for flooding local area.
Straw	Place straw on top of concrete.	Cheap and easy to apply.	Does not stay in place very well and can discolor concrete.
Plastic	Cover surface of concrete with sheets of plastic.	Cheap and easy to do	Can cause blotching of concrete.
Curing Compounds	Apply chemical with hand sprayer.	Do not have any work after initial installation.	More expensive and may need to be removed before applying floor covering.

CURLING OF CONCRETE SLABS

The curling of a concrete slab on grade is the result of the concrete drying faster on the top of the slab as compared to the bottom. The length of the top of the slab will shorten but the bottom of the slab will not cooperate with this movement since the ground will restrain it. This condition results in the corners of the slab curling upward.

The amount the corner of the slab lifts off from the ground is a function of the rate of shrinkage, the length between joints and the thickness of the slab. The following is a discussion of the importance of each of these items:

- Shrinkage Rate:

    Reduction of the rate at which moisture is leaving the top surface of the slab is a very effective way to reduce the difference in shrinkage from the top of the slab to the bottom. This can be accomplished by using proper curing techniques.

- Length between Joints

    The farther the distance between the control joints the greater the curling. In fact a reduction in the distance between the control joints by 1/2 will reduce the curling to 1/4 the previous amount.

- Slab Thickness

    The amount of curling is inversely proportional to the slab thickness. A slab that is 8 inches thick will have less than 1/2 the curling of a slab that is 4 inches thick.

- Others

    Wire mesh reinforcing located in the top third of the slab thickness is effective in reducing the effects of curling. A low slump concrete with reasonable good strength is the first step in preventing curling.

CONSOLIDATION OF CONCRETE

GENERAL INFORMATION

- Freshly poured concrete has entrapped air that results in discontinuities of the concrete and porous concrete.

- Freshly poured concrete must be consolidated to remove the trapped air.

- Vibration is the most commonly used technique for consolidating concrete.

- Vibration is accomplished by placing a steel shaft in the concrete. The shaft has a mass rotating inside that causes harmonic motion. The harmonic motion liquifies the concrete in the area of the vibrator and fills the air voids with concrete.

GENERAL PROCEDURES RECOMMENDED BY AMERICAN CONCRETE INSTITUTE (ACI 309)

- Deposit layers of concrete in 12 to 18 inch lifts.

- Level surface of pour.

- Insert vibrator vertically with uniform spacing.

- New area of vibrations should just overlap previously vibrated areas.

- Penetrate vibrator 6 inches into underlying layer.

- Hold stationary for 5 to 15 seconds.

- Withdraw vibrator at rate of 1 foot per four seconds.

Problems Associated with Too Large of a Vibrator

- Small air voids on surface
- Leaking at form joints
- Movement at form joints

Problems Associated with Too Small of a Vibrator

- Honeycombing

Problems Associated with Not Enough Vibration

- Honeycombing
- Pour lines visible in wall

Problems Associated with Over Vibration

- Sand Streaking

CONCRETE DURING EXTREME TEMPERATURES

HOT WEATHER (GUIDELINES PROVIDED BY
AMERICAN CONCRETE INSTITUTE ACI 305)

• Place concrete upon its arrival.

• Have backup vibrators since vibrators are more likely to break down in hot weather.

• Temperature will most likely impact concrete when temperature is above 75 degrees.

• Be prepared to promptly provide proper curing.

• The addition of cold water will lower concrete mix temperatures.

• Do not use cements or aggregates that have high temperatures.

• Do not schedule pours in the hottest part of the day.

• Try to delay pours until slabs are protected from direct sunlight by walls and other members.

COLD WEATHER (GUIDELINES PROVIDED BY
AMERICAN CONCRETE INSTITUTE ACI 306)

• Cold weather is defined as three days of mean daily temperatures below 40 degrees F.

• For residential construction concrete shall be maintained at not less than 55 degrees F.

• Concrete should not be placed on frozen subgrade.

• All snow and ice must be removed from the forms.

• Avoid water curing in cold weather.

• Foundations should have heat protection for two days.

- Concrete can take about 25% to 100% longer to gain strength in cold weather.

- Leave forms in place during weather protection period.

TESTING OF FRESH CONCRETE

SAMPLING FRESHLY MIXED CONCRETE (ASTM C172)

- Take sample as soon as possible.

- Start slump test within five minutes.

- Test two or more samples.

- Sampling shall take care to have samples truly representative of material.

- Only take samples after all water has been added.

- Sampling time shall not exceed 15 minutes.

- Start air content within 5 minutes.

CURING OF FIELD TEST SPECIMENS (ASTM C31)

- Molds shall be watertight and of nonabsorbent material.

- Compressive test specimens shall be cured with samples upright.

- Compressive test standard cylinder shall be 6 inches diameter by 12 inches high.

- Collect sample in three layers.

- Consolidate each layer with 25 strokes per layer. Tap outside of mold 10 to 15 times.

- Record location where sample was taken and time of sampling.

- Sample shall be initially covered with impervious plastic.

- When tests are needed to determine when concrete can be put in service, the cylinders shall be stored as close as possible to point of deposit.

- The cylinders shall be subjected to conditions as near to that of the structure as possible.

DRILLING CORES (ASTM C42)

- Concrete shall be 14 days old before cores are taken.

- Coring shall be perpendicular to bed of concrete as originally poured.

- Avoid having reinforcing bars in cores that are to be tested.

- Core should have a least a 4 inch diameter and have a length twice the diameter.

OTHER TESTS

- Air Content (ASTM C231, C138 or C173)

- Temperature Measurement (ASTM C1064)

- Cement and Water Content (ASTM C1078 and C1079)

- Unit Weight (ASTM C138)

- Rate of Hardening (ASTM C403)

- Density (ASTM C1040)

TESTING OF HARDENED CONCRETE

- Windsor Probe (ASTM C803):

    Estimates concrete strength. A steel probe is impacted into the concrete by a powder activated gun. The depth of penetration indicates the approximate concrete strength.

- Pullout Test (ASTM C900):

    Estimates concrete strength. A steel rod is embedded in the concrete when it is placed. The rod is pulled out and the corresponding force determined. The pull out strength provides an approximate measurement of the compressive strength.

- Rebound Hammer (ASTM C805):

    Estimates concrete strength. A spring loaded piston is struck against the concrete. The rebound of the piston provides an estimate of the compressive strength of the concrete.

JOINTS IN CONCRETE SLABS

Concrete shrinks after being placed, expands when heated, and contracts when cooled. This movement can produce stresses that cause cracking. Therefore, proper detailing of concrete members is necessary to prevent the buildup of stresses caused by this movement. The building up of stresses can be relieved by the use of joints in the concrete. There are three types of joints for concrete slabs:

• Construction Joints:

> The location where the concrete placement was stopped for more than 30 minutes or more.

• Isolation Joints:

> A gap between two concrete members, such as a column and slab that allows them to move independent of each other so that no cracking occurs.

• Control Joints:

> A groove in the surface of concrete to induce the concrete to crack at a selected location.

GENERAL GUIDELINES FOR CONTROL JOINT

• Joints shall be a maximum of 1/4 inch wide and shall have a depth equal to 1/4th the slab thickness.

• The maximum spacing of control joints in inches shall not exceed 30 times the slab thickness.

• If the plan width of a strip of concrete is greater than 10 feet, the use of a longitudinal control joint should be considered.

• The panels formed by control joints should be square. Panels with length to width ratios greater than 1.5 are not recommended.

• Joints shall be continuous and not be offset.

Table 5.2—Concrete Admixtures

Type	Standard	Purpose	Comments
Accelerators	ASTM D98	Provides a higher strength at an early age. Typical applications are when it is desired to have form work removed early, concrete must be loaded at an early age, or when traffic is to be allowed on pavement as soon as possible.	During placement operations the concrete may set up faster than desired. Chemicals in the admixture may cause premature deterioration of the steel reinforcing.
Air-entraining	ASTM C260	Air-entraining is the introduction of small air bubbles into the concrete mix. These air bubbles improve the durability of the concrete because water that freezes at the surface of the concrete can expand into the bubbles rather than scaling the surface.	Air-entraining improves the workability of concrete. However, too much air-entraining will result in a reduction of the concrete strength.
Pumpability		Admixtures that prevent the dewatering of the concrete under the pressure of pumping.	Pumpability admixtures will change the results of the mix design since it may increase the amount of water needed and cause air-entrainment.
Retarders	ASTM C494	Slows down the rate at which the concrete hardens. Typical application may be when high air temperatures decrease normal setting time or to delay setting time for the use of special surface finishes.	The retarders will cause a slower rate of strength production. Use may result in additional air-entrainment.
Water Reducers	ASTM C494	Reduces amount of water needed for a workable mixture yielding higher strength concrete.	Can result in increase in shrinkage of concrete.

TOLERANCES OF FORMED CONCRETE

Building Lines:

> The variation of building line from plan shall not exceed 1 inch.

Footings:

> Width of footing shall not be less than 1/2 inch than that specified.

> Width of footing shall not be greater than 2 inches than that specified.

Openings:

> Location and size of sleeve, wall and floor openings shall of variation of no more than 1/4 inch.

Member Sizes:

> The thickness, depth and width of slabs, beams, columns and walls shall not be less than 1/4 inch of that specified.

> The thickness, depth and width of slabs, beams, columns and walls shall not be greater than 1/2 inch of that specified.

Level:

> Lintels, horizontal grooves and sills shall not vary more than 1/2 inch for the entire length.

> Lintels, horizontal grooves and sills shall not vary more than 1/4 inch in any bay or within a 20 ft length.

Plumb:

> Lines and surfaces of columns and walls shall not vary by more than 1/4 inch in any 10 feet.

> Lines and surfaces of columns and walls shall not vary by more than 1 inch in its entire length.

Lines and surfaces of vertical lines shall not vary by more than 1/4 inch in any 10 feet.

Lines and surfaces of columns and walls shall not vary by more than 1/2 inch in its entire length.

TOLERANCES OF FLOOR SURFACES

The flatness of a floor can be estimated using the Face Floor Profile Number method, which is commonly referred to as the F-Numbers. The American Society for Testing and Materials (ASTM) has a standard for this procedure called "Standard Test Method for Determining Floor Flatness and Levelness Using the F-Number System." The purpose of the test is to obtain statistical information about the profile of a concrete floor. Elevations of the floor are measured at 12 inch intervals and then statistically analyzed to provide an estimate of the floor's flatness and levelness.

INTRODUCTION TO TILT-UP CONCRETE

Tilt-up concrete is a method of economically building concrete walls that began more than 50 years ago. A horizontal concrete slab is cast which later will be used as a wall. After the concrete has obtained the proper strength the concrete slab is tilted upwards to act as a wall. For purposes of this discussion, tilt-up includes walls built on site and in a precast plant.

This type of construction has gained such popularity that between 5000 and 8000 buildings are constructed using tilt-up each year. This system is used for all types of buildings but has gained a significant market share in one-story industrial buildings.

The reason for the use of tilt-up construction is that this type of wall requires little maintenance, has good fire and sound transmission resistance and can be constructed fairly quickly as compared to other types of construction.

The following is the sequence of construction for tilt-up walls that are constructed off site:

- Field work: Excavate for continuous foundation.

- Field work: Place continuous concrete foundation.

- Off site: Set up formwork in horizontal position.

- Off site: Place blockout, windows and lifting hardware into form.

- Off site: Place concrete in form works.

- Off site: Allow concrete to gain strength and then strip forms.

- Travel: Deliver panels to site.

- Field: Lift panels into position.

- Field: Connect panels together.

Note that the panels could have been formed adjacent to their final position and literally tipped into position.

Table 5.3—Finishing of Concrete Types

Type	Application	Details
Troweled	Slab	Surface shall be troweled but completed work shall be free of trowel marks.
Broom	Slab	Move broom across surface of concrete after float finish.
Dry Shake	Slab	Apply two-thirds of metallic aggregate and float finish surface. Apply remainder in perpendicular direction.
Exposed Aggregate	Slab	Spread exposed aggregate and lightly push into concrete surface. Remove concrete surrounding sides of aggregate.
Rough Form	Wall	Form work tie holes and defects to be repaired. Form work fins greater than 1/4 inch shall be removed.
Smooth Form	Wall	Material shall be placed on inside of form work to produce desired finish.
Smooth Rubbed	Wall	Within 24 hours of removal of forms, rub wall with abrasive material to get uniform color or texture.

Table 5.4—Finishing of Concrete Default Types

Use	Default
Floor for walking surface	Troweled
Ramps	Broom
Concrete surfaces not exposed to view of public	Rough form
Concrete surfaces exposed to public view	Smooth form

Table 5.5—Large Welded Wire Fabric
General Properties

Type	Spacing (in.)	Diameter	Square Inch of Area Per Foot
W8	8 × 8	0.32	0.12
W8	6 × 6	0.32	0.16
W8	4 × 4	0.32	0.24
W7	8 × 8	0.30	0.11
W7	6 × 6	0.30	0.14
W7	4 × 4	0.30	0.21
W6	8 × 8	0.28	0.09
W6	6 × 6	0.28	0.12
W6	4 × 4	0.28	0.18
W5	8 × 8	0.25	0.08
W5	6 × 6	0.25	0.10
W5	4 × 4	0.25	0.15
W4	8 × 8	0.23	0.06
W4	6 × 6	0.23	0.07
W4	4 × 4	0.23	0.16

Table 5.6—Standard Concrete Reinforcing Bar Information

Bar #	Diameter (in.)	Diameter (in.)	Cross-Sectional Area (in. sq.)	Weight Per Foot (lbs)
4	1/2	0.50	0.20	0.69
5	5/8	0.63	0.31	1.04
6	3/4	0.75	0.44	1.50
7	7/8	0.88	0.60	2.04
8	1	1.00	0.79	2.67
9	9/8	1.128	1.00	3.40

Table 5.7—Welded Wire Fabric
General Properties

Type	Spacing (in.)	Diameter	Square Inch of Area Per foot
W2.9	4 × 4	0.19	0.09
W2.9	6 × 6	0.19	0.06
W2	4 × 4	0.16	0.06
W2	6 × 6	0.16	0.04
W1.4	4 × 4	0.13	0.04
W1.4	6 × 6	0.13	0.03

NOTES:

NOTES:

NOTES:

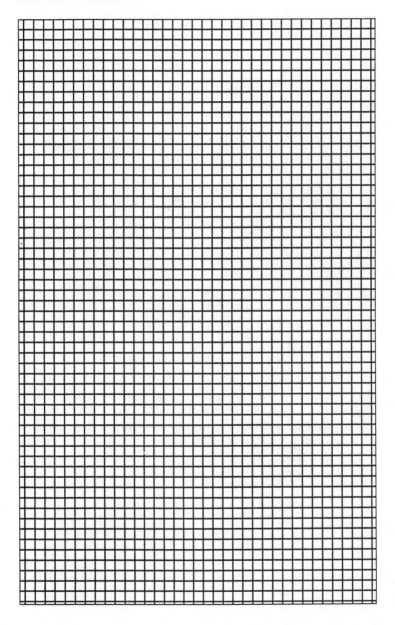

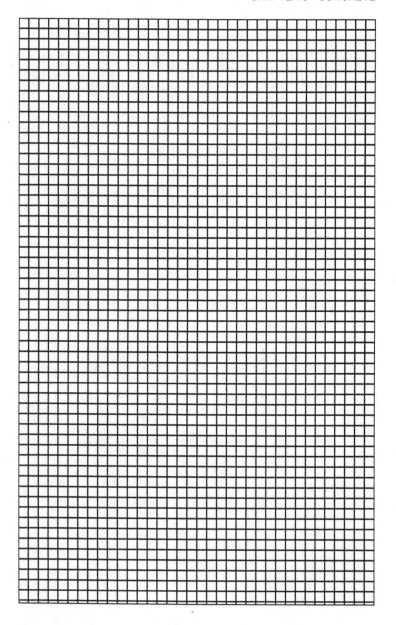

Chapter 6
Masonry

TIPS & INFORMATION

•When constructing a wall remove bricks from all stacks simulta-
neously to provide for uniform color. Variations in piles can be
blended together by this technique.

• Flashing at the base of the wall will deflect water that has entered
the wall cavity out to the exterior. Properly install flashing on
clean surfaces with proper overlaps at splices.

- Rowlock placement of bricks as window ledges or wall topping has proven to be troublesome for colder climates. This positioning of the brick allows for deterioration of the brick in a short time frame.

- Materials to be mixed to form mortar should be protected from moisture. The addition of saturated sand will introduce water in to the mix that has not been accounted for.

- Steel ties are essential for tying the brick wall to the wood or masonry framing. The ties should be properly spaced in both the vertical and horizontal directions.

- Avoid placing masonry in extreme temperatures. Mortar placed in hot weather and cold weather can cause shrinkage and unhydrated cement, respectively.

- When performing tuckpointing repairs, remove all loose mortar. Skim coating mortar joints does little to increase the life of the masonry.

MASONRY TOLERANCES

AMERICAN CONCRETE INSTITUTE SPECIFICATIONS (INCHES)

Bed Joint Thickness	+1/8 to -1/8
Head Joint Thickness	+3/8 to -1/4
Collar Joint Thickness	+3/8 to -1/4
Cavity Width	+3/8 to -1/4
Cross Section	+1/2 to -1/4
Levelness of Bed Joints	+1/4 to -1/4 in 10 ft
Levelness of Bed Joints Limit	+1/2 to -1/2 max
Levelness of Top of Bearing Wall	+1/4 to -1/4 in 10 ft
Levelness of Top of Bearing Wall Limit	+1/2 to -1/2 max
Plumbness	+1/4 to -1/4 in 10 ft
Alignment of top to bottom of bearing column	+1/2 to -1/2
Alignment of top to bottom of nonbearing column	+3/4 to -3/4
Alignment of top to bottom of bearing wall	+1/2 to -1/2
Alignment of top to bottom of nonbearing wall	+3/4 to -3/4

- 3/8" bed joint
- 15 ⅝ block length

- 3/8" head joint
- 7 ⅝ block height

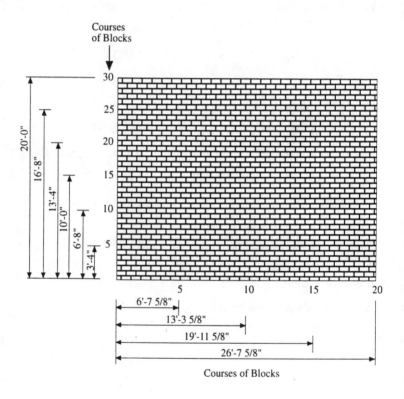

Figure 6.1—Modular Masonry Layout

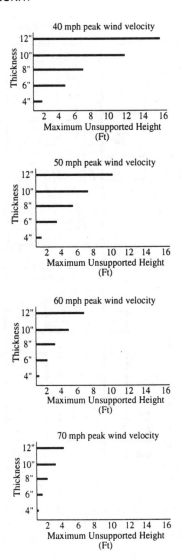

Figure 6.2—Unsupported Heights of Masonry Walls

Double Basket Weave   Triple Basket Weave

Stepped Basket Weave   Mixed Basket Weave

Figure 6.3—Masonry Patterns

Diagonal Bond   Basket Weave

Diagonal Basket Weave   Double Diagonal Basket Weave

Figure 6.4—Masonry Patterns (cont.)

Running Bond

Horizontal Stacking

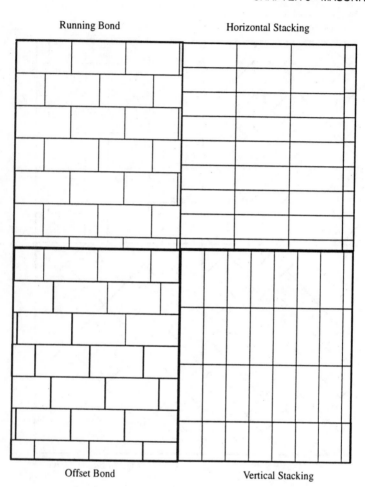

Offset Bond

Vertical Stacking

Figure 6.5—Masonry Patterns

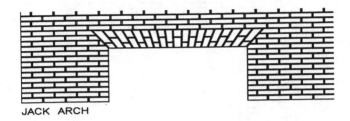

JACK ARCH

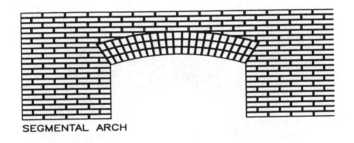

SEGMENTAL ARCH

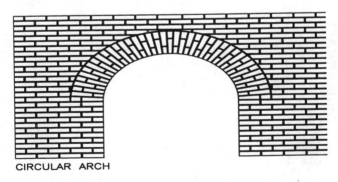

CIRCULAR ARCH

Figure 6.6—Types of arches

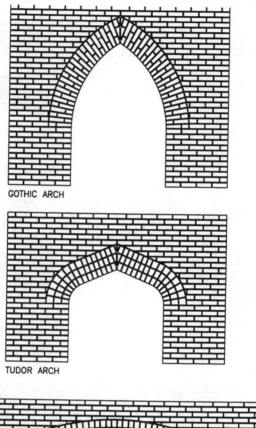

GOTHIC ARCH

TUDOR ARCH

PARABOLIC ARCH

Figure 6.7—Types of Arches (cont.)

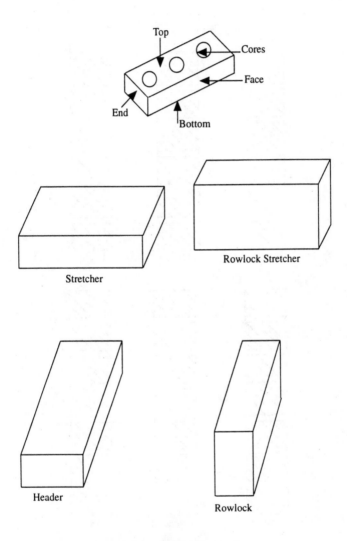

Figure 6.8—Brick Nomenclature

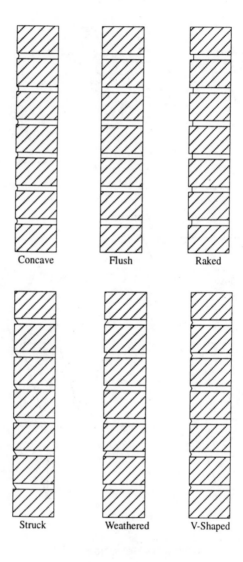

Figure 6.9—Mortar Joints

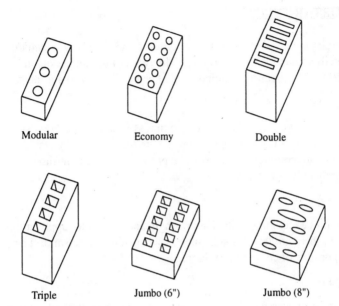

Modular   Economy   Double

Triple   Jumbo (6")   Jumbo (8")

Name	Thickness (inch)	Length (inch)	Height (inch)
Modular	4	2 ⅔	8
Economy	4	4	8
Double	4	5 ⅓	8
Triple	4	5 ⅓	12
6" jumbo	6	4	12
8" Jumbo	8	4	12

MORTAR TYPES

Mortars are divided into different types by ASTM C270, Standard Specifications for Unit Masonry. The types of mortar are distinguished by the ratio of components of cement, lime and aggregate.

Type M

This is the type with the greatest compressive strength (2500 psi at 28 days). It is used in applications where the wall is subjected to significant compressive loads. The proportion of cement/lime/aggregate is 1/0.25/3.5.

Type S

Provides excellent tensile bond strength and is used for applications where there are lateral loads, such as wind, soil, and seismic. The proportions of cement/lime/aggregate is 1/0.5/4.5.

Type N

Used for veneers and load bearing walls. The proportion of cement/lime/aggregate is 1/1/6.

Type O

Used in applications where freezing and thawing are not expected. The proportion of cement/lime/aggregate is 1/2/9.

TYPES OF MASONRY WALLS

CAVITY WALLS

- Facing wythes and backing wythes separated by air space.

- Connected by metal ties spaced 24 in. vertically and 36" horizontally as a maximum.

- Maximum height/thickness = 18.

- Minimum thickness = 8 in.

- Cavity space: 1 to 4 in.

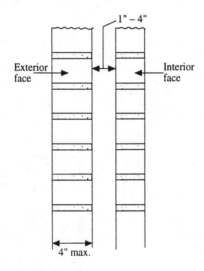

HOLLOW UNIT MASONRY

- Constructed of hollow masonry units.

- Typically refers to unreinforced concrete.

- Maximum height/thickness = 18.

- Minimum thickness = 8 in.

SOLID MASONRY

- Bonding of wythes can be by metal reinforcing or by a brick header.

- Distance between headers should not exceed 24 inches vertically or horizontally.

- Ends of metal ties must embed no less than 1" into adjacent wythe.

- If wythe space not filled, the wall will be treated as a cavity wall.

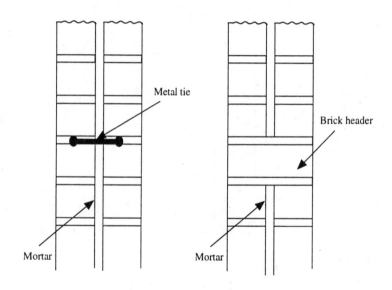

REINFORCED GROUTED MASONRY

- Two wythes of brick in which the interior joint is filled with grout.

- Grout between wythes to be at least 1/4 in. and not to exceed 3 in. depending on grout type.

- Joint between wythes is called a collar joint.

- Use type M or S mortar typically.

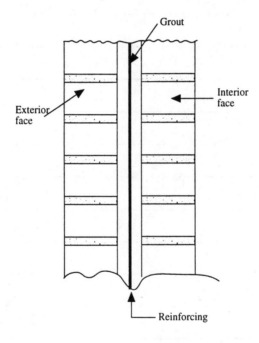

NOTES:

NOTES:

NOTES:

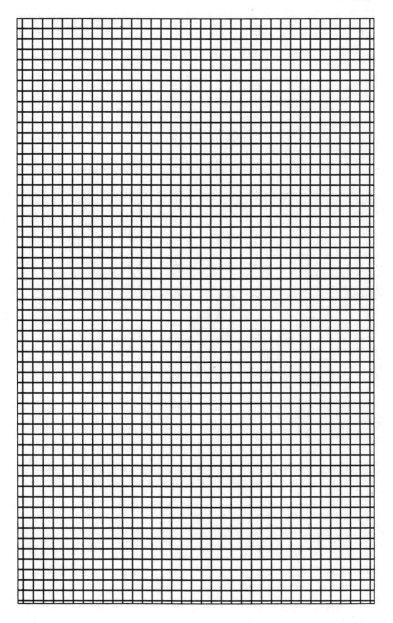

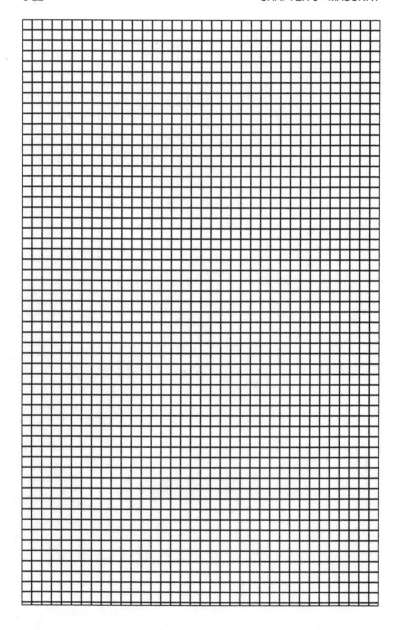

Chapter 7
Wood

TIPS & INFORMATION

- For typical framing the roof structure must form a triangle. To satisfy this criteria, the rafter ends must be connected to a horizontal ceiling joist. If this joist is not present, the horizontal thrust produced by roof loads will push the top of the walls outward.

- Prefabricated trusses are usually designed by sophisticated software. The truss members are held together by truss plates that are machine rolled into place.

- Roof trusses are typically unstable until the roof decking is installed. Temporary bracing must be installed to prevent tip over.

- Notches or holes in wood members should be done at selected locations. Holes should be drilled at midspan of members. Notches should never be placed at the bottom of a typical joist. The bottom of the joist is in tension and will tend to tear open at the notch.

- Building codes typically list nailing requirements. These are usually presented in tabular formats and provide information on nailing joists to sill plates, roof rafters to ridge boards, ceiling joists to plates, etc.

- The joist ends are typically covered by a band board that runs perpendicular to the joists. On the first floor it is possible that joists to build a deck may be hung from the band board. For this reason it is recommended that the band board not be constructed of plywood or a similar material.

- Collar ties resist lateral thrust transferred to the roof rafters. The higher the collar, ties are installed towards the peak, the greater the stresses in the roof members.

- Note that a ridge board is different than a ridge beam. A ridge board is a nailer for the upper ends of the roof rafters. A ridge beam is a structural member that must be capable of transferring loads to its end supports.

DESIGN COMPRESSION STRENGTH OF WOOD (PSI)

Excellent

Beech	1400–2000
Birch	1400–2000
Hickory	1800–2000
Hem Fir	1200–1900
Northern Red Oak	1300–1900
Redwood	1300–2400
Southern Pine	1400–2800

Very Good

Douglas Fir (North)	1100–1800
Douglas Fir (South)	1100–1800
Mixed Oak	1100–1600
Red Maple	1200–1800
Spruce-Pine-Fir	1200–1700
White Oak	1200–1600

Good

Aspen	800–1200
Eastern White Pines	800–1700
Mixed Maple	900–1400
Western Cedars	900–1400
Yellow Poplar	1000–1400

Average

Northern White Cedar	700–1000

DEFLECTION RESISTANCES OF WOOD (MODULUS OF ELASTICITY)

Excellent

Beech	1,500,000–1,700,000
Birch	1,500,000–1,700,000
Douglas Fir (North)	1,600,000–1,900,000
Hem Fir	1,300,000–1,600,000
Hickory	1,500,000–1,700,000
Red Maple	1,500,000–1,700,000
Southern Pine	1,600,000–1,900,000

Very Good

Douglas Fir (South)	1,200,000–1,400,000
Northern Red Oak	1,300,000–1,400,000
Redwood	1,100,000–1,400,000
Spruce-Pine-Fir	1,400,000–1,500,000
Yellow Poplar	1,300,000–1,500,000

Good

Aspen	900,000–1,100,000
Eastern White Pines	1,100,000–1,200,000
Mixed Maple	1,100,000–1,300,000
Mixed Oak	900,000–1,100,000
Western Cedar	1,000,000–1,100,000
White Oak	900,000–1,100,000

Average

Southern White Cedar	700,000–800,000

Table 7.1—Geometric Properties of Wood Members

Site	Width (t)	Depth (d)	Area (A)	Moment of Inertia (I)	Section Modulus (S)
1 × 3	3/4	2 1/2	1.88	0.98	0.78
1 × 4	3/4	3 1/2	2.63	2.68	1.53
1 × 6	3/4	5 1/2	4.13	10.40	3.78
2 × 4	1 1/2	3 1/2	5.25	5.36	30.6
2 × 6	1 1/2	5 1/2	8.25	20.80	7.56
2 × 8	1 1/2	7 1/4	10.88	47.64	13.14
2 × 10	1 1/2	9 1/4	13.88	98.93	21.39
2 × 12	1 1/2	11 1/4	16.88	117.98	31.64
2 × 14	1 1/2	13 1/4	19.88	290.78	43.89
3 × 4	2 1/2	3 1/2	8.75	8.93	5.10
4 × 4	3 1/2	3 1/2	12.25	12.51	7.15
4 × 6	3 1/2	5 1/2	19.25	48.53	17.65
6 × 6	5 1/2	5 1/2	30.25	76.26	27.75

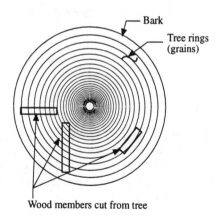

Figure 7.1—Wood members cut from tree with different grain orientation

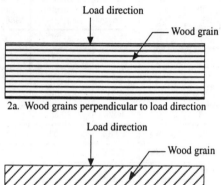

2a. Wood grains perpendicular to load direction

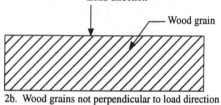

2b. Wood grains not perpendicular to load direction

Figure 7.2—Orientation of wood grains

OVERVIEW OF FACTORS THAT MODIFY WOOD JOIST STRENGTH

Load Duration

Effect: Short duration is better.

Amount: Results in increase of 100% for short duration. 15% to
              69% for medium duration loads.

Reason: Wood members can carry heavy loads if the loads are
              for a short duration. An example of a short duration
              load would be the impact forces of a dropped object.
              Loads that are more permanent in nature include floor
              live loads and dead loads.

Moisture

Effect: Reduction.

Amount: Decrease in strength of 3% to 15%.

Reason: When the moisture content is high (greater than 19%),
            the properties of wood become less desirable.

Stability Factor

Effect: Reduction.

Amount: As much as 90% of the bending allowable.

Reason: Lateral support must be provided to prevent joists from
            buckling sideways. If the joist are supported laterally at
            the top by decking and bridging, the beam stability
            factor is 1.0. The reduction has a greater value for the
            less support provided.

Beam Size Factor

Effect: Usually an increase.

Amount: 2×4 (50%), 2×6 (30%), 2×8 (20%), 2×10 (10%),
            2×12 (10% reduction).

Reason: Allowable stresses provided for wood members is
            provided considering 2×12 members. If a wood
            member is smaller than a 2×12, a greater allowable
            stress is allowed.

Repetitive Member Factor

Effect: Increase.

Amount: 15%.

Reason: When wood is used for members such as floor joists and
            ceiling joists, the members are placed side by side.
            Because of this redundancy an increase in the stress is
            allowed if three members involved are spaced no
            more than 24 inches apart.

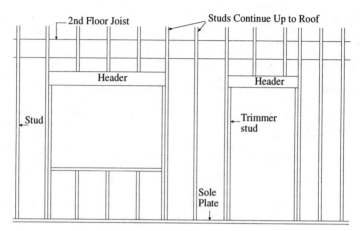

Figure 7.3—Balloon framing

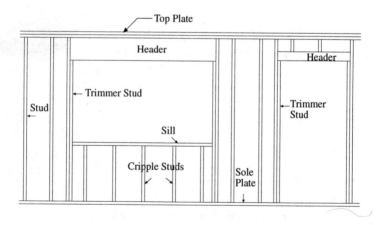

Figure 7.4—Platform framing

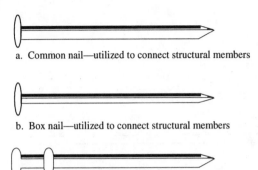

a. Common nail—utilized to connect structural members

b. Box nail—utilized to connect structural members

c. Double-headed nail—used in an operation where the nail is only temporary and is later removed. Common for nailing formwork.

d. Finish nail—used when the head of the nail is not to be visible. Common for nailing interior trim.

e. Roofing nail—used to fasten shingles to the wood roof sheathing.

f. Drywall nail—used to attach drywall to the wall studs and ceiling

Figure 7.5—Nails used in residential construction

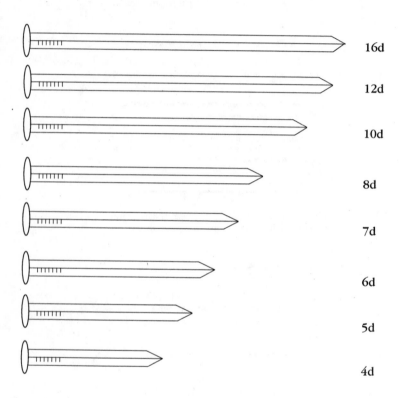

Figure 7.6—Nail actual lengths

Table 7.2—Common Nail Sizes

Size	Length (in.)	Diameter (in.)
4d	1.50	0.10
5d	1.75	0.10
6d	2.00	0.11
7d	2.25	0.11
8d	2.50	0.13
10d	3.00	0.14
12d	3.25	0.14
16d	3.50	0.16

Table 7.3—Box Nail Sizes

Size	Length (in.)	Diameter (in.)
4d	1.50	0.08
5d	1.75	0.08
6d	2.00	0.10
7d	2.25	0.10
8d	2.50	0.11
10d	3.00	0.12
12d	3.25	0.12
16d	3.50	0.13

Table 7.4—Withdraw Strength of Nails in Side Grain
(pounds per inch of penetration)

Species / Density	Nail Type	8d	10d	12d	16d	20d
Western Cedars	Common Nail	13	15	15	16	19
.35	Box Nail	11	13	13	14	15
Balsam Fir	Common Nail	14	16	16	17	21
.36	Box Nail	12	14	14	14	16
Aspen	Common Nail	18	21	21	23	27
.40	Box Nail	16	18	18	14	21
Eastern Spruce	Common Nail	19	22	22	24	29
.41	Box Nail	17	19	19	20	22
Hem-Fir	Common Nail	21	23	23	26	30
.42	Box Nail	18	20	20	21	23
Spruce-Pine-fir	Common Nail	21	23	23	26	30
.42	Box Nail	18	20	20	21	23
Yellow Poplar	Common Nail	26	29	29	32	38
.46	Box Nail	22	25	25	27	29
Ponderosa Pine	Common Nail	30	34	34	38	45
.49	Box Nail	26	30	30	31	34
Southern Pine	Common Nail	41	46	46	50	59
.55	Box Nail	35	40	40	42	46

Table 7.5—Minimum Penetration Depth for
Lateral Load Design Values
(inches of penetration into member holding nail point)

Species / Density	Nail Type	8d	10d	12d	16d	20d
Western Cedars	Common Nail	1 3/4	2	2	2 1/4	2 3/4
	Box Nail	1 5/8	1 3/4	1 3/4	1 3/4	2
Balsam Fir	Common Nail	1 3/4	2	2	2 1/4	2 3/4
	Box Nail	1 5/8	1 3/4	1 3/4	1 3/4	2
Aspen	Common Nail	1 3/4	2	2	2 1/4	2 3/4
	Box Nail	1 5/8	1 3/4	1 3/4	1 3/4	2
Eastern Spruce	Common Nail	1 3/4	2	2	2 1/4	2 3/4
	Box Nail	1 5/8	1 3/4	1 3/4	1 3/4	2
Hem Fir	Common Nail	1 3/4	2	2	2 1/4	2 1/2
	Box Nail	1 1/2	1 3/4	1 3/4	1 3/4	2
Spruce-Pine Fir	Common Nail	1 3/4	2	2	2 1/4	2 1/2
	Box Nail	1 1/2	1 3/4	1 3/4	1 3/4	2
Yellow Poplar	Common Nail	1 3/4	2	2	2 1/4	2 1/2
	Box Nail	1 1/2	1 3/4	1 3/4	1 3/4	2
Ponderosa Pine	Common Nail	1 3/4	2	2	2 1/4	2 1/2
	Box Nail	1 1/2	1 3/4	1 3/4	1 3/4	2
Southern Pine	Common Nail	1 1/2	1 5/8	1 5/8	1 3/4	2 1/4
	Box Nail	1 1/4	1 1/2	1 1/2	1 1/2	1 3/4

Table 7.6—Allowable Lateral Load for Nails in Single Shear

Species \ Density	Nail Type	8d	10d	12d	16d	20d
Western Cedars	Common Nail	51	61	61	70	91
	Box Nail	41	49	49	54	61
Balsam Fir	Common Nail	51	61	61	70	91
	Box Nail	41	49	49	54	61
Aspen	Common Nail	51	61	61	70	91
	Box Nail	41	49	49	54	61
Eastern Spruce	Common Nail	51	61	61	70	91
	Box Nail	41	49	49	54	61
Hem Fir	Common Nail	64	77	77	88	91
	Box Nail	51	62	62	67	61
Spruce-Pine Fir	Common Nail	64	77	77	88	114
	Box Nail	51	62	62	67	77
Yellow Poplar	Common Nail	64	77	77	88	114
	Box Nail	51	62	62	67	77
Ponderosa Pine	Common Nail	64	77	77	88	114
	Box Nail	51	62	62	67	77
Southern Pine	Common Nail	78	94	94	108	139
	Box Nail	63	76	76	82	94

Table 7.7—Nailing Schedule for Structural Members
(using common or box nails)

Members to be Nailed	Member Nailed to	Nail Size	# of Nails	Spacing	Nailing Type
Joist	Sill Plate	8 penny	3	n/a	Toe nail
Joist	Girder	8 penny	3	n/a	Toe nail
Top Plate	Stud	16 penny	2	n/a	End nail
Double Top Plate	Top Plate	10 penny	n/a	24 in.	Face nail
Double Studs	Stud	10 penny	n/a	24 in.	Face nail
Builtup Header	Header	16 penny	n/a	16 in.	Edges
Ceiling Joist	Plate	8 penny	3	n/a	Toe nail
Ceiling Joist Partition Lap	Ceiling Joist	10 penny	3	n/a	Face nail
Ceiling Joists	Parallel Rafters	10 penny	3	n/a	Face nail
Rafter	Plate	16 penny	2	n/a	Toe nail
Roof Rafter	Ridge	16 penny	4	n/a	Toe nail
Roof Rafter	Hip Rafters	16 penny	4	n/a	Toe nail
Rafter Ties	Rafters	8 penny	3	n/a	Face nail

Table 7.8—Plywood Nailing Schedule
(using common nails & wind speeds less than 80 mph)

Members to be Nailed	Member Nailed to	Nail Size	# of Nails	Spacing	Nailing Type
1/2" thick plywood (edges)	Floor Joists	6 penny	n/a	6 in.	Straight
3/4" thick plywood (edges)	Floor Joists	8 penny	n/a	6 in.	Straight
1/2" thick plywood (edges)	Walls	6 penny	n/a	6 in	Straight
3/4" thick plywood (edges)	Walls	8 penny	n/a	6 in	Straight
1/2" thick plywood (edges)	Rafters	8 penny	n/a	6 in	Straight
3/4" thick plywood (edges)	Rafters	8 penny	n/a	6 in	Straight
1/2" thick plywood (intermediate)	Floor Joists	6 penny	n/a	12 in.	Straight
3/4" thick plywood (intermediate)	Floor Joists	8 penny	n/a	12 in.	Straight
1/2" thick plywood (intermediate)	Walls	6 penny	n/a	12 in.	Straight
3/4" thick plywood (intermediate)	Walls	8 penny	n/a	12 in.	Straight
1/2" thick plywood (intermediate)	Rafters	8 penny	n/a	12 in.	Straight
3/4" thick plywood (intermediate)	Rafters	8 penny	n/a	12 in.	Straight

HEM-FIR

Members: • California Red Fir   • Grand Fir

• Noble Fir   • Pacific Noble Fir

• Western Hemlock   • White Fir

Source: Grown in northwestern region of the United States

Strength Characteristics: Excellent

Deflection Resistance: Excellent

Decay Resistance: Slightly or nonresistant

Nailing Strength: Good

Bending Strength Design Values (psi):

	Select Structural	No. 1	No. 2
2 × 4	2415	1640	1465
2 × 6	2095	1420	1270
2 × 8	1930	1310	1175
2 × 10	1770	1200	1075
2 × 12	1610	1095	980

Modulus of Elasticity: Select Structural–1,600,000 psi

No. 1–1,500,000

No. 2–1,300,000

Horizontal Shear: 75 psi

FLOOR JOIST—HEM-FIR
16 inches on Center/Live Load = 30 psf/Dead Load = 10 psf

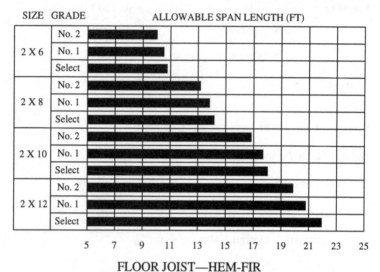

FLOOR JOIST—HEM-FIR
16 inches on Center/Live Load = 40 psf/Dead Load = 10 psf

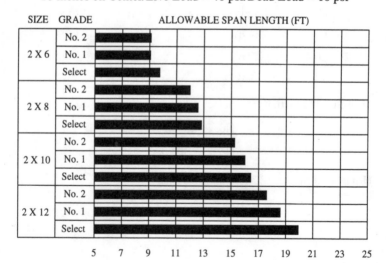

CEILING JOIST—HEM-FIR
16 inches on Center/Live Load = 10 psf/Dead Load = 5 psf

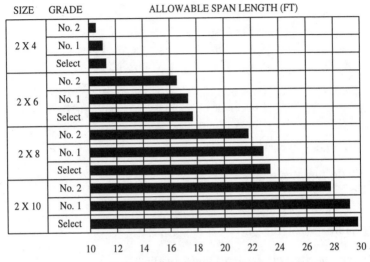

CEILING JOIST—HEM-FIR
16 inches on Center/Live Load = 20 psf/Dead Load = 10 psf

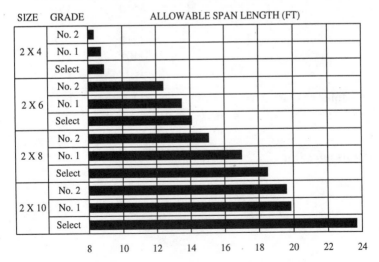

<u>S-P-F</u>
<u>(SPRUCE-PINE-FIR)</u>

Members:
- Alpine
- Black Spruce
- Jack Pine
- Red Spruce
- Balsam Fir
- Englemann Spruce
- Lodgepole Pine
- White Spruce

Source: Grown in east coast and northern midwest states. Lodgepole pine and Englemann Spruce are from Rocky Mountain areas.

Strength Characteristics: Very good

Deflection Resistance: Very good

Decay Resistance: Slightly or nonresistant

Nailing Strength: Good

Bending Strength Design Values (psi):

	Select Structural	No. 1	No.2
2 × 4	2155	1510	1510
2 × 6	1870	1310	1310
2 × 8	1725	1210	1210
2 × 10	1580	1105	1105
2 × 12	1440	1005	1005

Modulus of Elasticity: Select Structural–1,500,000 psi
No. 1–1,400,000
No. 2–1,400,000

Horizontal Shear: 70 psi

FLOOR JOIST—SPRUCE-PINE-FIR
16 inches on Center/Live Load = 30 psf/Dead Load = 10 psf

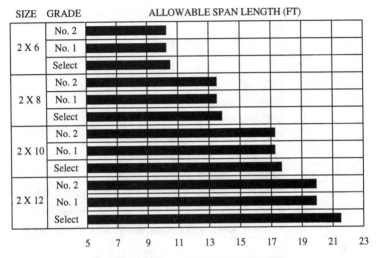

FLOOR JOIST—SPRUCE-PINE-FIR
16 inches on Center/Live Load = 40 psf/Dead Load = 10 psf

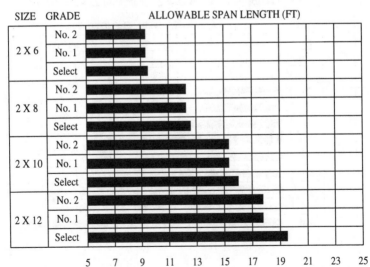

CEILING JOIST—SPRUCE-PINE-FIR
16 inches on Center/Live Load = 10 psf/Dead Load = 5 psf

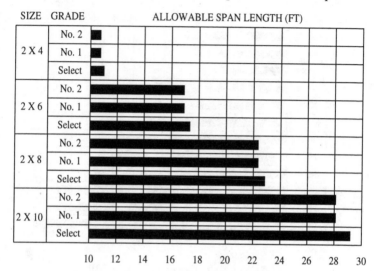

CEILING JOIST—SPRUCE-PINE-FIR
16 inches on Center/Live Load = 20 psf/Dead Load = 10 psf

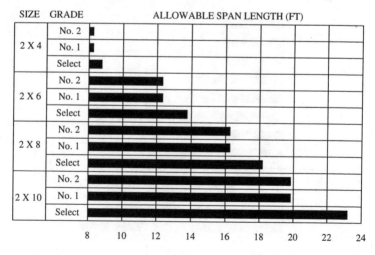

RAFTER—SPRUCE-PINE-FIR/16 inches on Center
Slope greater than 3/12/Maximum Deflection less than L/180
Live Load = 20 psf/Dead Load = 15 psf

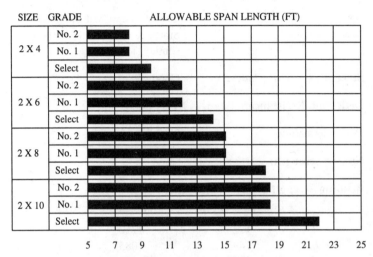

RAFTER—SPRUCE-PINE-FIR/16 inches on Center
Slope greater than 3/12/Maximum Deflection less than L/180
Live Load = 40 psf/Dead Load = 15 psf

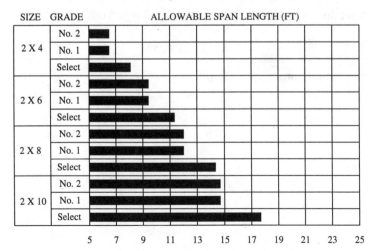

YELLOW POPLAR

Members: Yellow Poplar only

Source: Grown in east half of United
                                States predominantly in the
                                southeast.

Strength Characteristics: Good

Deflection Resistance: Very good

Decay Resistance: Slightly or nonresistant

Nailing Strength: Good

Bending Strength Design Values (psi):

	Select Structural	No. 1	No. 2
2 × 4	1725	1250	1210
2 × 6	1495	1085	1045
2 × 8	1380	1000	965
2 × 10	1265	915	885
2 × 12	1150	835	805

Modulus of Elasticity: Select Structural–1,500,000 psi
                             No. 1–1,400,000
                             No. 2–1,300,000

Horizontal Shear: 75 psi

FLOOR JOIST—YELLOW POPLAR
16 inches on Center/Live Load = 30 psf/Dead Load = 10 psf

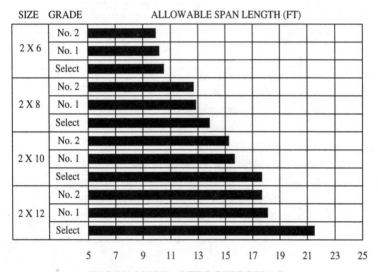

FLOOR JOIST—YELLOW POPLAR
16 inches on Center/Live Load = 40 psf/Dead Load = 10 psf

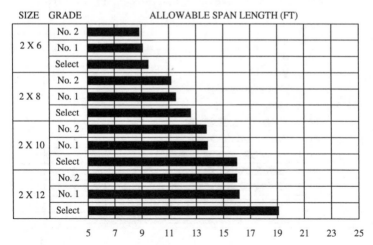

CEILING JOIST—YELLOW POPLAR
16 inches on Center/Live Load = 10 psf/Dead Load = 5 psf

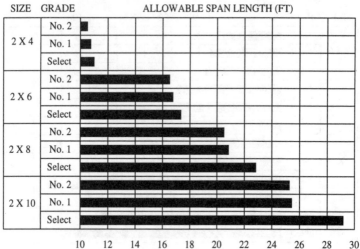

CEILING JOISTS—YELLOW POPLAR
16 inches on Center/Live Load = 20 psf/Dead Load = 10 psf

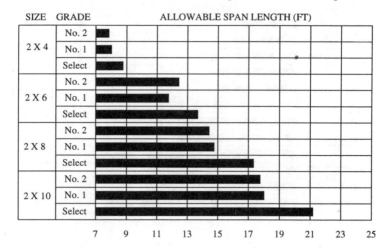

RAFTER—YELLOW POPLAR/16 inches on Center
Slope greater than 3/12/Maximum Deflection less than L/180/Live
Load = 20 psf/Dead Load = 15 psf

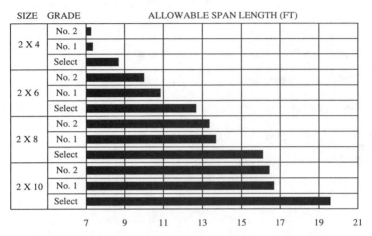

RAFTER—YELLOW POPLAR/16 inches on Center
Slope greater than 3/12/Maximum Deflection less than L/180/Live
Load = 40 psf/Dead Load = 15 psf

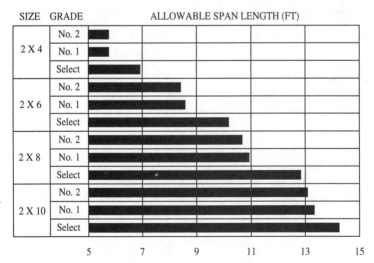

<u>RED MAPLE</u>

Member: Red Maple only

Source: Northern midwestern states

Strength Characteristics: Very good

Deflection Resistance: Excellent

Decay Resistance: Slightly or nonresistant

Nailing Strength: Excellent

Bending Strength Design Values (psi):

	<u>Select Structural</u>	<u>No. 1</u>	<u>No. 2</u>
2 × 4	2245	1595	1555
2 × 6	1945	1385	1345
2 × 8	1795	1275	1240
2 × 10	1645	1170	1140
2 × 12	1495	1065	1035

Modulus of Elasticity: Select Structural–1,700,000 psi
                              No. 1–1,600,000
                              No. 2–1,500,000

Horizontal Shear: 105 psi

FLOOR JOIST—RED MAPLE
16 inches on Center/Live Load = 30 psf/Dead Load = 10 psf

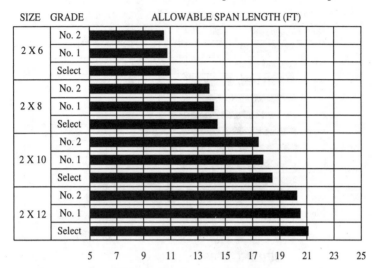

FLOOR JOIST—RED MAPLE
16 inches on Center/Live Load = 40 psf/Dead Load = 10 psf

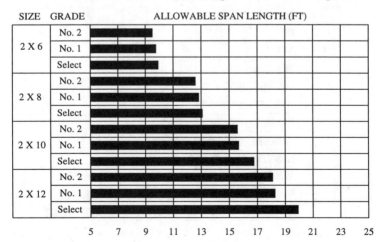

CEILING JOIST—RED MAPLE
16 inches on Center/Live Load = 20 psf/Dead Load = 10 psf

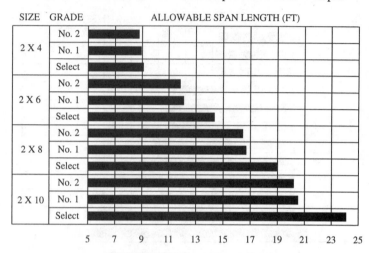

SIZE   GRADE   ALLOWABLE SPAN LENGTH (FT)

CEILING JOIST—RED MAPLE
16 inches on Center/Live Load = 10 psf/Dead Load = 5 psf

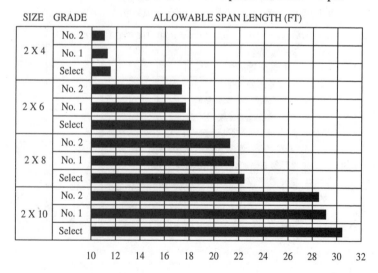

SIZE   GRADE   ALLOWABLE SPAN LENGTH (FT)

RAFTER—RED MAPLE/16 inches on Center
Slope greater than 3/12/Maximum Deflection less than L/180
Live Load = 20 psf/Dead Load = 15 psf

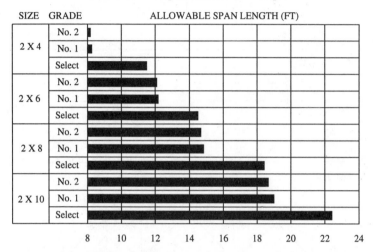

RAFTER—RED MAPLE/16 inches on Center
Slope greater than 3/12/Maximum Deflection less than L/180
Live Load = 40 psf/Dead Load = 15 psf

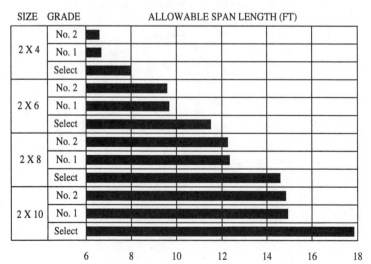

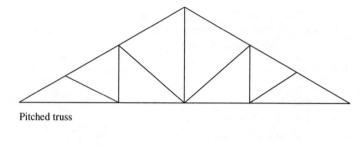

Pitched truss

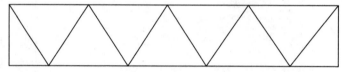

Flat truss

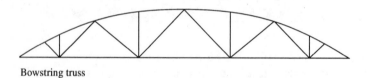

Bowstring truss

Figure 7.7—Truss designation by shape

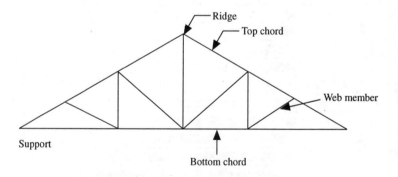

Figure 7.8—Components of a truss

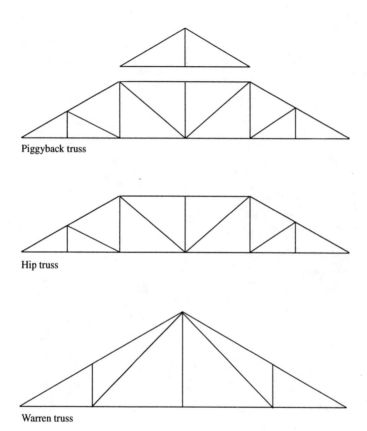

Piggyback truss

Hip truss

Warren truss

Figure 7.9—Types of trusses

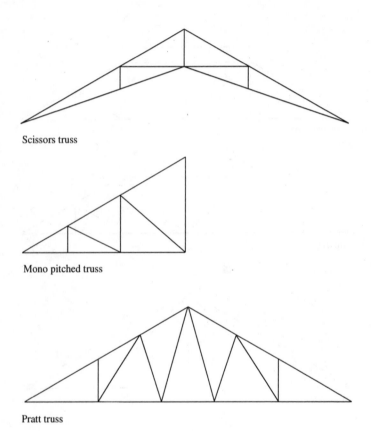

Scissors truss

Mono pitched truss

Pratt truss

Types of trusses (cont.)

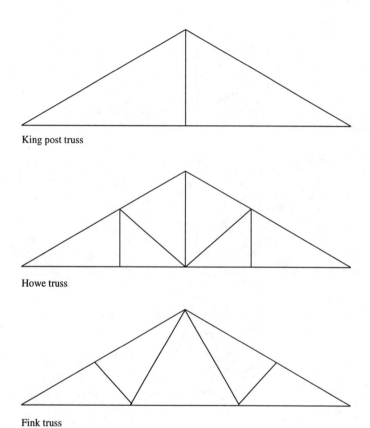

King post truss

Howe truss

Fink truss

Types of trusses (cont.)

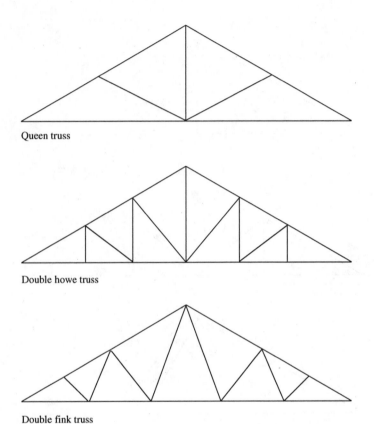

Queen truss

Double howe truss

Double fink truss

Types of trusses (cont.)

BRACING OF TRUSSES DURING ERECTION

Wood truss systems are an economical method for spanning lengths from 10 feet to 40 feet in residential construction and up to 80 feet in commercial construction. Their widespread use can be attributed as much to ease of installation as to cost effectiveness. The majority of trusses used in construction are typically pre-fabricated.

Despite their wide use there appears to be a limited knowledge by field personnel about the proper installation of a pre-fabricated wood truss. During installation, trusses are highly susceptible to tipping over. A collapse of a single truss can cause a domino effect that could cause the failure of an entire roof system and lead to the injury of employees.

Temporary bracing is needed to prevent the tipping of trusses. The bracing is used to stabilize the trusses until the roof decking is installed. A contractor installing wood trusses should be thoroughly familiar with the publication "Temporary Bracing of Metal Plate Connected Wood Trusses" published by the Truss Plate Institute (TPI). The Truss Plate Institute is headquartered in Madison, Wisconsin and provides a wealth of information on truss installation.

The following pages present information on the temporary bracing of trusses. This information is based on the work presented in the TPI document but is basically presented in a less intimidating fashion. Information presented herein is not a substitute for that publication, and it is essential that the reader add that publication to their library of essential books and understand the contents.

TPI recommends a temporary bracing system that consists of several different types of braces. The braces that will be discussed here will include the bracing that is present within the trusses themselves (i.e., no ground bracing). The following four types of braces are required according to TPI:

TOP CHORD LATERAL BRACING

Lateral bracing is installed on the top side of the trusses, orientated perpendicular to the truss span direction and is continuous from end to end. The function of this bracing is to reduce the unbraced length of the top chords. The bracing must be a least 2 × 4's in minimum lengths of 10 feet. The tables that follow these introductory pages provide the number of lateral bracing lines needed. The following illustration depicts a layout of top chord lateral bracing.

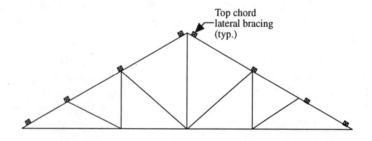

Top chord
lateral bracing
(typ.)

BOTTOM CHORD LATERAL BRACING

Used to stabilize the bottom chord during installation and to maintain proper spacing between trusses. Most installations place the bottom chord temporary lateral bracing at the location where the permanent bracing is required. Bottom chord bracing shall not be farther apart than 15 feet from adjacent bottom chord lateral bracing. The following illustration depicts a layout of bottom chord lateral bracing.

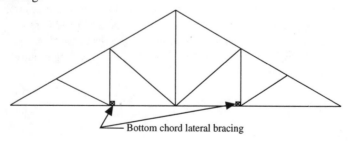

Bottom chord lateral bracing

TOP CHORD DIAGONAL BRACES

Top chord diagonals are required between the top chord lateral bracing as shown in the following figure. These braces are installed at 45 degree angles and shall be at least 2 × 4's, installed with at least two 16 penny nails. The tables that follow this introductory discussion provide the maximum distance that the top chord diagonal braces can be spaced.

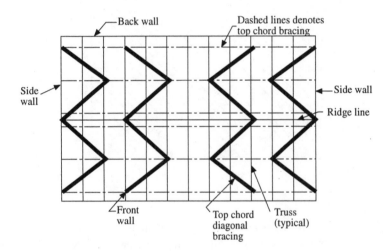

WEB PLAN BRACING

Web plane bracing is formed by attaching diagonal bracing on opposite sides of the same group of similar web members to form an "X." The bracing shall be repeated at no greater than 20 foot intervals. The following illustration depicts a layout of web plane bracing.

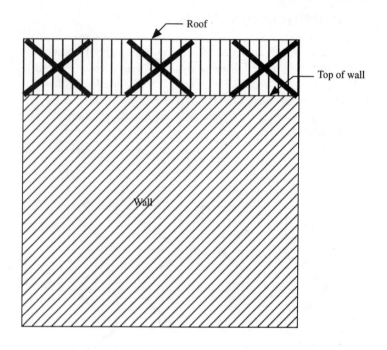

The tables that follow provide a suggested amount and location of bracing for trusses during erection. The tables have information for both dual pitched trusses and flat trusses. The suggested bracing is not the minimum required but, instead, provides conservative bracing suggestions. In many cases less bracing can be installed and be in conformance with codes and recommended practices. The suggestions presented provide an efficient and conservative method of bracing the trusses.

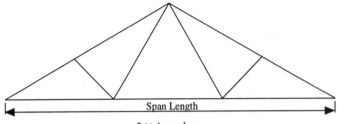

2 × 4 members
4 top chord panels

Pitch	Spacing (ft)	Type of Bracing	Number or Spacing					
2/12	4	No. of top chord lateral	6	8				
		Top chord diagonal spacing	24	16				
2/12	2	No. of top chord lateral	6		8			
		Top chord diagonal spacing	18		14			
3/12	4	No. of top chord lateral	6	8				
		Top chord diagonal spacing	32	24				
3/12	2	No. of top chord lateral	6		8			
		Top chord diagonal spacing	26		18			
4/12	4	No. of top chord lateral	6	8				
		Top chord diagonal spacing	40	28				
4/12	2	No. of top chord lateral	6		8			
		Top chord diagonal spacing	32		22			
5/12	4	No. of top chord lateral	6	8				
		Top chord diagonal spacing	48	36				
5/12	2	No. of top chord lateral	6		8			
		Top chord diagonal spacing	36		26			
6/12	4	No. of top chord lateral	6	8				
		Top chord diagonal spacing	52	40				
6/12	2	No. of top chord lateral	6		8			
		Top chord diagonal spacing	42		30			

16   24   32   40   48   56

Span Length
(ft)

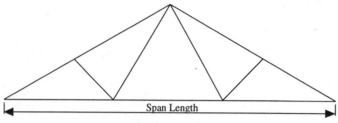

Span Length

2 × 8 members
4 top chord panels

Pitch	Spacing (ft)	Type of Bracing	Number or Spacing				
2/12	4	No. of top chord lateral	6	8			
		Top chord diagonal spacing	16	12			
2/12	2	No. of top chord lateral	6		8		
		Top chord diagonal spacing	14		10		
3/12	4	No. of top chord lateral	6	8			
		Top chord diagonal spacing	24	16			
3/12	2	No. of top chord lateral	6		8		
		Top chord diagonal spacing	20		12		
4/12	4	No. of top chord lateral	6	8			
		Top chord diagonal spacing	32	20			
4/12	2	No. of top chord lateral	6		8		
		Top chord diagonal spacing	24		16		
5/12	4	No. of top chord lateral	6	8			
		Top chord diagonal spacing	36	24			
5/12	2	No. of top chord lateral	6		8		
		Top chord diagonal spacing	36		18		
6/12	4	No. of top chord lateral	6	8			
		Top chord diagonal spacing	40	28			
6/12	2	No. of top chord lateral	6		8		
		Top chord diagonal spacing	32		22		

16   24   32   40   48   56

Span Length
(ft)

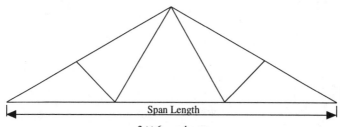

2 × 6 members
4 top chord panels

Pitch	Spacing (ft)	Type of Bracing	Number or Spacing					
2/12	4	No. of top chord lateral	6	8				
		Top chord diagonal spacing	20	12				
2/12	2	No. of top chord lateral	6		8			
		Top chord diagonal spacing	16		10			
3/12	4	No. of top chord lateral	6	8				
		Top chord diagonal spacing	28	20				
3/12	2	No. of top chord lateral	6		8			
		Top chord diagonal spacing	22		14			
4/12	4	No. of top chord lateral	6	8				
		Top chord diagonal spacing	32	24				
4/12	2	No. of top chord lateral	6		8			
		Top chord diagonal spacing	26		18			
5/12	4	No. of top chord lateral	6	8				
		Top chord diagonal spacing	40	28				
5/12	2	No. of top chord lateral	6		8			
		Top chord diagonal spacing	30		20			
6/12	4	No. of top chord lateral	6	8				
		Top chord diagonal spacing	44	32				
6/12	2	No. of top chord lateral	6		8			
		Top chord diagonal spacing	34		24			

16   24   32   40   48   56

Span Length
(ft)

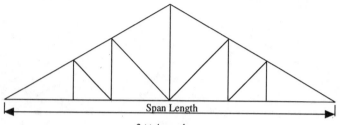

2 × 4 members
6 top chord panels

Pitch	Spacing (ft)	Type of Bracing	Number or Spacing					
2/12	4	No. of top chord lateral	6	8				
		Top chord diagonal spacing	24	16				
2/12	2	No. of top chord lateral	6		8			
		Top chord diagonal spacing	18		12			
3/12	4	No. of top chord lateral	6	8				
		Top chord diagonal spacing	32	20				
3/12	2	No. of top chord lateral	6		8			
		Top chord diagonal spacing	24		16			
4/12	4	No. of top chord lateral	6	8				
		Top chord diagonal spacing	36	28				
4/12	2	No. of top chord lateral	6		8			
		Top chord diagonal spacing	30		18			
5/12	4	No. of top chord lateral	6	8				
		Top chord diagonal spacing	44	32				
5/12	2	No. of top chord lateral	6		8			
		Top chord diagonal spacing	36		22			
6/12	4	No. of top chord lateral	6	8				
		Top chord diagonal spacing	52	36				
6/12	2	No. of top chord lateral	6		8			
		Top chord diagonal spacing	40		24			

16   24   32   40   48   56

Span Length
(ft)

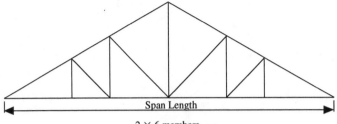

2 × 6 members
6 top chord panels

Pitch	Spacing (ft)	Type of Bracing	Number or Spacing				
2/12	4	No. of top chord lateral	6	8			
		Top chord diagonal spacing	20	12			
2/12	2	No. of top chord lateral	6		8		
		Top chord diagonal spacing	16		10		
3/12	4	No. of top chord lateral	6		8		
		Top chord diagonal spacing	24		16		
3/12	2	No. of top chord lateral	6			8	
		Top chord diagonal spacing	22			10	
4/12	4	No. of top chord lateral	6		8		
		Top chord diagonal spacing	32		20		
4/12	2	No. of top chord lateral	6			8	
		Top chord diagonal spacing	26			14	
5/12	4	No. of top chord lateral	6		8		
		Top chord diagonal spacing	36		24		
5/12	2	No. of top chord lateral	6			8	
		Top chord diagonal spacing	28			16	
6/12	4	No. of top chord lateral	6		8		
		Top chord diagonal spacing	44		28		
6/12	2	No. of top chord lateral	6			8	
		Top chord diagonal spacing	32			18	

16   24   32   40   48   56

Span Length
(ft)

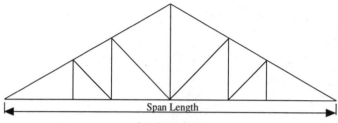

2 × 8 members
6 top chord panels

Pitch	Spacing (ft)	Type of Bracing	Number or Spacing					
2/12	4	No. of top chord lateral	6	8				
		Top chord diagonal spacing	16	12				
2/12	2	No. of top chord lateral	6		8			
		Top chord diagonal spacing	14		8			
3/12	4	No. of top chord lateral	6	8				
		Top chord diagonal spacing	24	16				
3/12	2	No. of top chord lateral	6		8			
		Top chord diagonal spacing	18		10			
4/12	4	No. of top chord lateral	6		8			
		Top chord diagonal spacing	28		16			
4/12	2	No. of top chord lateral	6			8		
		Top chord diagonal spacing	22			14		
5/12	4	No. of top chord lateral	6		8			
		Top chord diagonal spacing	36		20			
5/12	2	No. of top chord lateral	6			8		
		Top chord diagonal spacing	26			16		
6/12	4	No. of top chord lateral	6		8			
		Top chord diagonal spacing	40		24			
6/12	2	No. of top chord lateral	6			8		
		Top chord diagonal spacing	30			18		
			16	24	32	40	48	56

Span Length
(ft)

2 × 4 TRUSS MEMBERS
TRUSS SPACING = 2 FEET

Number of Lateral Braces Needed

Span Length / Truss Spacing	24"	30"	36"	40"	48"
20 feet	3	3	3	3	3
25 feet	4	4	4	4	4
30 feet	6	5	5	5	5
35 feet	7	7	6	6	6
40 feet	10	9	8	8	7
45 feet	–	–	11	10	9

Maximum Spacing of Diagonal Braces

Span Length / Truss Spacing	24"	30"	36"	40"	48"
20 feet	48	60	74	88	100
25 feet	26	28	34	32	36
30 feet	16	20	22	28	30
35 feet	12	12	16	18	24
40 feet	6	8	12	12	16
45 feet	–	–	8	10	12

2 × 4 TRUSS MEMBERS
TRUSS SPACING = 4 FEET

Number of Lateral Braces Needed

Span Length / Truss Spacing	24"	30"	36"	40"	48"
20 feet	4	4	4	4	4
25 feet	5	5	5	4	4
30 feet	7	7	6	6	6
35 feet	10	9	8	8	8
40 feet	–	–	–	11	10
45 feet	–	–	–	–	–

Maximum Spacing of Diagonal Braces

Span Length / Truss Spacing	24"	30"	36"	40"	48"
20 feet	36	44	48	92	100
25 feet	28	32	40	52	52
30 feet	16	20	24	28	36
35 feet	12	12	16	20	20
40 feet	–	–	–	12	16
45 feet	–	–	–	–	–

2 × 6 TRUSS MEMBERS
TRUSS SPACING = 4 FEET

Number of Lateral Braces Needed

Span Length / Truss Spacing	24"	30"	36"	40"	48"
20 feet	4	3	3	3	3
25 feet	5	4	4	4	4
30 feet	6	6	5	5	5
35 feet	9	8	7	7	6
40 feet	–	10	9	9	8
45 feet	–	–	–	11	10

Maximum Spacing of Diagonal Braces

Span Length / Truss Spacing	24"	30"	36"	40"	48"
20 feet	28	60	68	84	96
25 feet	20	32	36	36	40
30 feet	16	16	24	28	28
35 feet	8	12	16	16	20
40 feet	–	8	12	12	16
45 feet	–	–	8	8	16

2 × 6 TRUSS MEMBERS
TRUSS SPACING = 2 FEET

Number of Lateral Braces Needed

Span Length / Truss Spacing	24"	30"	36"	40"	48"
20 feet	3	4	4	4	4
25 feet	4	4	4	4	4
30 feet	5	4	4	4	4
35 feet	6	6	5	5	5
40 feet	8	7	7	6	6
45 feet	–	10	9	9	9

Maximum Spacing of Diagonal Braces

Span Length / Truss Spacing	24"	30"	36"	40"	48"
20 feet	42	56	68	78	88
25 feet	18	20	24	24	26
30 feet	12	20	12	24	26
35 feet	10	12	16	18	18
40 feet	6	8	10	14	14
45 feet	–	6	6	8	10

2 × 8 TRUSS MEMBERS
TRUSS SPACING = 2 FEET

Number of Lateral Braces Needed

Span Length / Truss Spacing	24"	30"	36"	40"	48"
20 feet	4	4	4	4	4
25 feet	4	4	4	4	4
30 feet	5	4	4	4	4
35 feet	6	5	5	5	5
40 feet	8	7	6	6	6
45 feet	10	9	8	8	7

Maximum Spacing of Diagonal Braces

Span Length / Truss Spacing	24"	30"	36"	40"	48"
20 feet	16	18	20	22	22
25 feet	16	18	20	22	22
30 feet	10	18	20	22	18
35 feet	8	12	14	14	22
40 feet	6	8	10	12	12
45 feet	4	4	6	8	10

2 × 8 TRUSS MEMBERS
TRUSS SPACING = 4 FEET

Number of Lateral braces Needed

Span Length / Truss Spacing	24"	30"	36"	40"	48"
20 feet	4	3	3	3	3
25 feet	4	4	4	4	4
30 feet	6	5	5	5	4
35 feet	8	7	7	6	6
40 feet	11	10	8	8	7
45 feet	–	–	11	10	9

Maximum Spacing of Diagonal Braces

Span Length / Truss Spacing	24"	30"	36"	40"	48"
20 feet	24	52	60	80	92
25 feet	24	28	28	32	36
30 feet	12	20	24	24	36
35 feet	8	12	12	16	16
40 feet	4	8	8	12	12
45 feet	–	–	4	8	8

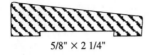

5/8" × 2 1/4"

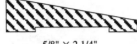

5/8" × 2 1/4"

11/16" × 2 1/2"

11/16" × 3 1/4"

9/16" × 3 1/4"

9/16" × 3 1/4"

1" × 4 1/2"

Figure 7.10—Casing Profiles (not to scale)

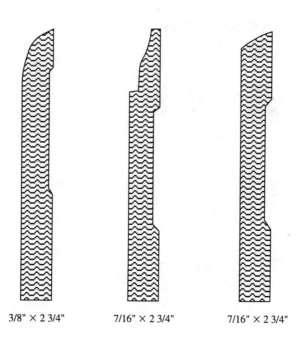

3/8" × 2 3/4"   7/16" × 2 3/4"   7/16" × 2 3/4"

Figure 7.11—Base Profiles

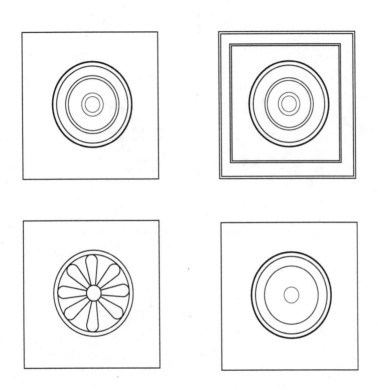

Figure 7.12—Rosettes

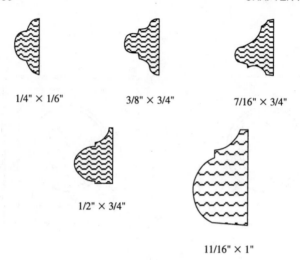

1/4" × 1/6"   3/8" × 3/4"   7/16" × 3/4"

1/2" × 3/4"

11/16" × 1"

Figure 7.13—Panel Mouldings (not to scale)

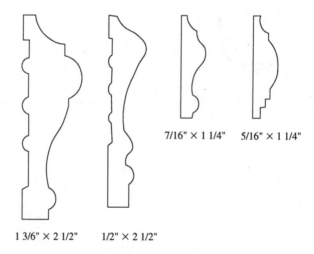

7/16" × 1 1/4"   5/16" × 1 1/4"

1 3/6" × 2 1/2"   1/2" × 2 1/2"

Figure 7.14—Chair Rail (not to scale)

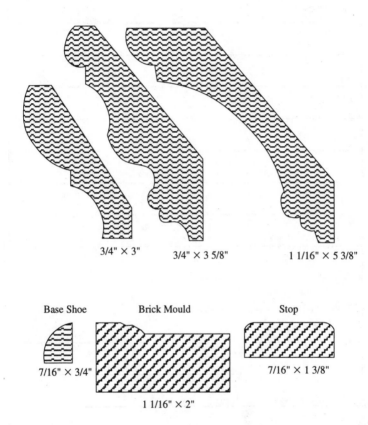

3/4" × 3"   3/4" × 3 5/8"   1 1/16" × 5 3/8"

Base Shoe   Brick Mould   Stop

7/16" × 3/4"   7/16" × 1 3/8"

1 1/16" × 2"

Figure 7.15—Crown Mouldings (not to scale)

NOTES:

NOTES:

NOTES:

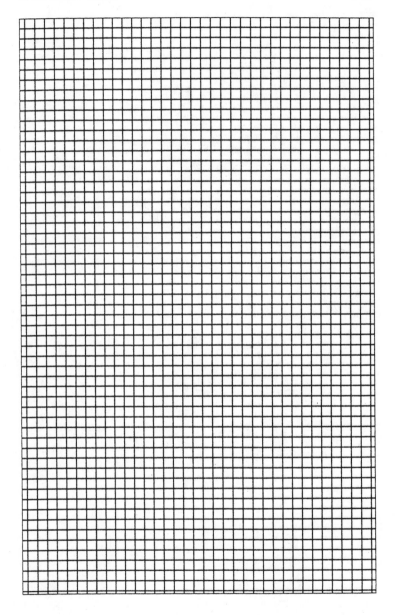

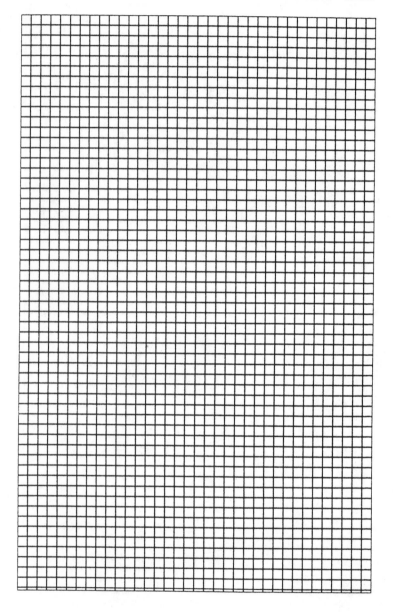

Chapter 8
Steel

TIPS & INFORMATION

- The cost of steel beams is relatively cheap. Greatly oversizing steel beams literally costs only a couple hundred extra dollars but provides a very stiff beam system. Avoid skimping on beam sizes.

- Check the depth of the steel beams prior to forming the beam pockets in the concrete formwork. If the concrete subcontractor blocks out for an eight inch deep, someone will have to jack hammer out two more inches if the plans call for a 10-inch deep beam.

- Plan out the layout for the basement for future use prior to finishing the plans for the house. Layout your basement steel to satisfy this future work. This will avoid a steel column located where it conflicts with the future basement use.

- In typical construction, the steel beams are not continuous and, instead, span in discrete segments (i.e., from wall to steel post where another segment begins). Consider using a continuous beam that is continuous from wall to wall or spans across several posts. This allows for the possibility of relocating posts in the future without having to change the steel beam.

- The steel beam will extend below the bottom of the joist, thereby encroaching on the basement headroom. Consider running the utilities, such as the heating ducts, so that a single soffit can be used for the beams and utilities.

- Coordinate the construction crew to install the beams when they are delivered. This coordination will allow the setting of the beams while the truck mounted crane is on site rather than hand setting later.

- Take care in setting the steel to have level beams in both directions.

The following provides a list of the most common steel beams used in residential construction with the corresponding geometrical properties. The designation of the steel beam provides the depth and weight of the member. For example, a W10 × 26 is approximately 10 inches deep and 26 pounds per foot of length. The geometrical property "I" provides a general idea of the load carrying capacity of the member. The larger the value of "I," the stronger the more load the member can carry.

Table 8.1—Typical Steel Members Used in Residential Construction

Designation	Area $(in.^2)$	Depth (in.)	Web Thickness (in.)	I $(in.^4)$	Flange Width (in.)	Flange Thickness (in.)
W10 × 26	7.61	10.33	0.260	144	5.77	0.44
W10 × 22	6.49	10.17	0.240	118	5.75	0.36
W10 × 19	5.62	10.24	0.250	96	4.02	0.39
W10 × 17	4.99	10.11	0.240	82	4.01	0.33
W8 × 24	7.08	7.93	0.245	83	6.50	0.40
W8 × 21	6.16	8.28	0.250	75	5.27	0.40
W8 × 18	5.26	8.14	0.230	62	5.25	0.33
W8 × 15	4.44	8.11	0.245	48	4.01	0.31
W8 × 13	3.84	7.99	0.230	40	4.00	0.25

The following table provides the allowable loads that can be carried by a steel beam and still be in compliance with building code requirements. Note that the table can only be used when uniform distributed loads are applied.

Table 8.2—Steel Beam Distributed Load Capacities (lb/ft)

Size		Length (ft)										
		8	9	10	11	12	13	14	15	16	17	18
W10 × 26	Total	6905	5456	4419	3652	3069	2615	2254	1964	1726	1529	1364
	Live	6905	5456	4419	3652	3069	2615	2254	1833	1510	1259	1060
W10 × 22	Total	5741	4536	3674	3036	2552	2174	1874	1633	1435	1271	1134
	Live	5741	4536	3674	3036	2552	2174	1874	1502	1237	1031	869
W10 × 19	Total	4652	3676	2977	2460	2068	1762	1518	1323	1163	1030	919
	Live	4652	3676	2977	2460	2068	1762	1507	1225	1010	842	709
W10 × 17	Total	4009	3168	2565	2120	1782	1518	1308	1140	1002	887	792
	Live	4009	3168	2565	2120	1782	1518	1308	1031	849	708	596
W8 × 24	Total	5172	4087	3310	2735	2299	1958	1688	1471	1292	1145	1021
	Live	5172	4087	3310	2673	2059	1619	1296	1054	868	724	610

Table 8.2—Steel Beam Distributed Load Capacities (lb/ft) (cont.)

W8 × 21	Total	4503	3559	2882	2381	2002	1705	1469	1280	1125	997	889
	Live	4503	3559	2882	2381	1872	1473	1179	959	790	658	555
W8 × 18	Total	3761	2972	2407	1989	1672	1424	1227	1069	940	833	743
	Live	3761	2972	2407	1989	1539	1210	969	787	649	541	456
W8 × 15	Total	2920	2307	1868	1544	1298	1105	953	830	729	646	576
	Live	2920	2307	1868	1544	1193	939	751	611	503	419	353
W8 × 13	Total	2452	1937	1569	1297	1090	928	800	697	613	543	484
	Live	2452	1937	1569	1278	984	774	620	504	415	346	291

A lintel is a structural member that will carry the load over an opening to the adjacent sides. The load is then carried down from the sides of the openings to the foundation. An ideal structural member for a lintel is a steel angle as shown in the figure below. Table 8.3 provides geometric properties of steel angles. Tables 8.4 and 8.5 provide allowable height of brick that can be carried by a given angle for a designated span.

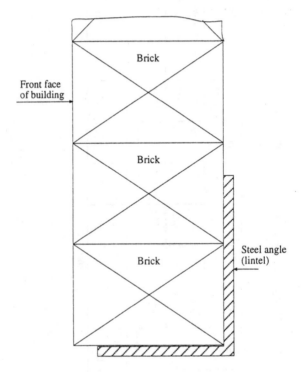

Figure 8.1—Lintel supporting brick

Table 8.3—Steel Angle Geometric Properties

Angle Designation	Vertical Leg (in.)	Horizontal Leg (in.)	Thickness (in.)	Weight per Foot (pounds)	I (in.^{4})	S (in.^{3})
3 × 3-1/2 × 1/4	3	3-1/2	1/4	5.4	1.30	0.59
3 × 3-1/2 × 5/16	3	3-1/2	5/16	6.6	1.58	0.72
3 × 3-1/2 × 3/8	3	3-1/2	3/8	7.9	1.85	0.85
3-1/2 × 3-1/2 × 1/4	3-1/2	3-1/2	1/4	5.8	2.01	0.79
3-1/2 × 3-1/2 × 5/16	3-1/2	3-1/2	5/16	7.2	2.45	0.98
3-1/2 × 3-1/2 × 3/8	3-1/2	3-1/2	3/8	8.5	2.87	1.15
4 × 3-1/2 × 1/4	4	3-1/2	1/4	6.2	2.91	1.03
4 × 3-1/2 × 5/16	4	3-1/2	5/16	7.7	3.56	1.26
4 × 4 × 1/4	4	4	1/4	6.6	3.04	1.05
4 × 4 × 5/16	4	4	5/16	8.2	3.71	1.29
4 × 4 × 3/8	4	4	3/8	9.8	4.36	1.52
4 × 4 × 1/2	4	4	1/2	12.8	5.56	1.97
5 × 3-1/2 × 5/16	5	3-1/2	5/16	8.7	6.60	1.94
5 × 3-1/2 × 3/8	5	3-1/2	3/8	10.4	7.78	2.29
5 × 3-1/2 × 1/2	5	3-1/2	1/2	13.6	9.99	2.99
6 × 3-1/2 × 5/16	6	3-1/2	5/16	9.8	10.9	2.73
6 × 3-1/2 × 3/8	6	3-1/2	3/8	11.7	12.9	3.24
6 × 4 × 3/8	6	4	3/8	12.3	13.5	3.32
6 × 4 × 1/2	6	4	1/2	16.2	17.4	4.33

Table 8.4—Allowable Height of Brick That Can Be Supported by
Steel Angle—Short Spans

Angle Size	Span ft					
	3	4	5	6	7	8
3 × 3-1/2 × 1/4	*	13.0	6.7	3.9	2.4	1.6
3 × 3-1/2 × 5/16	*	15.9	8.1	4.7	3.0	2.0
3 × 3-1/2 × 3/8	*	*	9.5	5.5	3.5	2.3
3-1/2 × 3-1/2 × 1/4	*	*	10.4	6.0	3.8	2.5
3-1/2 × 3-1/2 × 5/16	*	*	12.6	7.3	4.6	3.1
3-1/2 × 3-1/2 × 3/8	*	*	14.8	8.6	5.4	3.6
4 × 3-1/2 × 1/4	*	*	15.0	8.7	5.5	3.7
4 × 3-1/2 × 5/16	*	*	*	10.6	6.7	4.5
4 × 4 × 1/4	*	*	15.7	9.1	5.7	3.8
4 × 4 × 5/16	*	*	*	11.1	7.0	4.7
4 × 4 × 3/8	*	*	*	13.0	8.2	5.5
4 × 4 × 1/2	*	*	*	16.6	10.4	7.0

*Denotes height in excess of 18 ft

Table 8.5—Allowable Height of Brick Than Can Be Supported by Steel Angle—Long Spans

Angle Size	Span ft					
	9	10	11	12	13	14
3-1/2 × 3-1/2 × 3/8	2.5	1.8	1.4	0	0	0
4 × 3-1/2 × 1/4	2.6	1.9	1.4	0	0	0
4 × 3-1/2 × 5/16	3.1	2.3	1.7	1.3	0	0
4 × 4 × 1/4	2.7	2.0	1.5	1.1	0	0
4 × 4 × 5/16	3.3	2.4	1.8	1.4	1.1	0
4 × 4 × 3/8	3.9	2.8	2.1	1.6	1.3	0
4 × 4 × 1/2	4.9	3.6	2.7	2.1	1.6	1.3
5 × 3-1/2 × 5/16	5.8	4.3	3.2	2.5	1.9	1.6
5 × 3-1/2 × 3/8	6.9	5.0	3.8	2.9	2.3	1.8
5 × 3-1/2 × 1/2	8.8	6.4	4.8	3.7	2.9	2.3
6 × 3-1/2 × 5/16	9.6	7.0	5.3	4.1	3.2	2.6
6 × 3-1/2 × 3/8	11.4	8.3	6.2	4.8	3.8	3.0
6 × 4 × 3/8	11.9	8.7	6.5	5.0	4.0	3.2
6 × 4 × 1/2	15.4	11.2	8.4	6.5	5.1	4.1

If the brick walls extend uninterrupted upwards and to the sides, the lintel need only carry the weight of the brick shown in the triangles of Figure 8.2. The weight of the brick above the triangle is not carried by the lintels, rather it is carried by arching of the brick. The arching of the brick will transmit the loads outside the triangle to the sides of the opening.

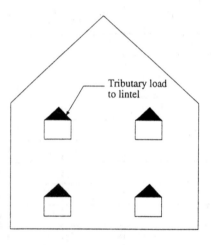

Figure 8.2—Shaded portion depicts the tributary area that is carried by each lintel as the result of arching

The benefit of arching can only be taken advantage of when the brick extends uninterrupted above and to the sides of the opening. It is generally agreed that the brick must extend upward a distance of one-half the opening plus another 1.5 ft. For an 8-ft opening the brick must extend upward a distance of 5.5 ft (8/2 + 1.5 = 5.5).

The length the brick must also extend to the side of the opening to have the benefit or arching action. The uninterrupted length the brick must extend is difficult to determine. A side length of one-half the opening plus an additional 0.5 ft is used in calculations in this text. It is believed that this is a reasonable distance.

Table 8.6—Special Conditions: Allowable Height of Brick That Can Be Supported by Steel Angle When Arching is Available—Short Spans

Angle Size	Span ft					
	3 See "A" below	4 See "B" below	5 See "C" below	6 See "D" below	7 See "E" below	8 See "F" below
3 × 3-1/2 × 1/4	No Limit	No Limit	No Limit	No Limit	2.3	1.5
3 × 3-1/2 × 5/16	No Limit	No Limit	No Limit	No Limit	2.8	1.8
3 × 3-1/2 × 3/8	No Limit	No Limit	No Limit	No Limit	No Limit	2.1
3-1/2 × 3-1/2 × 1/4	No Limit	No Limit	No Limit	No Limit	No Limit	2.3
3-1/2 × 3-1/2 × 5/16	No Limit	No Limit	No Limit	No Limit	No Limit	2.9
3-1/2 × 3-1/2 × 3/8	No Limit	No Limit	No Limit	No Limit	No Limit	3.4

Table 8.6—Special Conditions: Allowable Height of Brick That Can Be Supported by Steel Angle When Arching is Available—Short Spans (cont.)

4 × 3-1/2 × 1/4	No Limit	No Limit	No Limit	No Limit	No Limit	No Limit	No Limit
4 × 3-1/2 × 5/16	No Limit	No Limit	No Limit	No Limit	No Limit	No Limit	No Limit
4 × 4 × 1/4	No Limit	No Limit	No Limit	No Limit	No Limit	No Limit	No Limit
4 × 4 × 5/16	No Limit	No Limit	No Limit	No Limit	No Limit	No Limit	No Limit
4 × 4 × 3/8	No Limit	No Limit	No Limit	No Limit	No Limit	No Limit	No Limit
4 × 4 × 1/2	No Limit	No Limit	No Limit	No Limit	No Limit	No Limit	No Limit

"A"—Numbers are only valid if there is 3 ft of uninterrupted brick above lintel and 2 ft of uninterrupted brick on both sides of lintel

"B"—Numbers are only valid if there is 3 ½ ft of uninterrupted brick above lintel and 2 ½ ft of uninterrupted brick on both sides of lintel

"C"—Numbers are only valid if there is 4 ft of uninterrupted brick above lintel and 3 ft of uninterrupted brick on both sides of lintel

"D"—Numbers are only valid if there is 4 ½ ft of uninterrupted brick above lintel and 3 ½ ft of uninterrupted brick on both sides of lintel

"E"—Numbers are only valid if there is 5 ft of uninterrupted brick above lintel and 4 ft of uninterrupted brick on both sides of lintel

"F"—Numbers are only valid if there is 5 ½ ft of uninterrupted brick above lintel and 4 ½ ft of uninterrupted brick on both sides of lintel

Table 8.8—Special Conditions: Allowable Height of Brick That Can Be Supported By Steel Angle—Long Spans

| Angle Size | Span ft | | | | | |
	9 See "A" below	10 See "B" below	11 See "C" below	12 See "D" below	13 See "E" below	14 See "F" below
3-1/2 × 3-1/2 × 3/8	2.3	1.6	1.0	0	0	0
4 × 3-1/2 × 1/4	2.4	1.7	1.2	0	0	0
4 × 3-1/2 × 5/16	2.9	2.1	1.5	1.1	0	0
4 × 4 × 1/4	2.5	1.8	1.3	0	0	0
4 × 4 × 5/16	3.0	2.1	1.5	1.1	0	0
4 × 4 × 3/8	3.6	2.5	1.8	1.3	1.0	0
4 × 4 × 1/2	No Limit	3.2	2.3	1.7	1.3	0

Table 8.8—Special Conditions: Allowable Height of Brick That Can Be Supported By Steel Angle—Long Spans (cont.)

	"A"	"B"	"C"	"D"	"E"	"F"
5 × 3-1/2 × 5/16	No Limit	4.0	2.9	2.2	1.7	1.3
5 × 3-1/2 × 3/8	No Limit	No Limit	3.5	2.6	2.0	1.5
5 × 3-1/2 × 1/2	No Limit	No Limit	No Limit	3.3	2.5	2.0
6 × 3-1/2 × 5/16	No Limit	No Limit	No Limit	3.8	2.9	2.3
6 × 3-1/2 × 3/8	No Limit	No Limit	No Limit	4.5	3.4	2.7
6 × 4 × 3/8	No Limit	No Limit	No Limit	4.7	3.5	2.8
6 × 4 × 1/2	No Limit	No Limit	No Limit	No Limit	4.6	3.6

"A"—Numbers are only valid if there is 6 ft of uninterrupted brick above lintel and 5 ft of uninterrupted brick on both sides of lintel

"B"—Numbers are only valid if there is 6 ½ ft of uninterrupted brick above lintel and 5 ½ ft of uninterrupted brick on both sides of lintel

"C"—Numbers are only valid if there is 7 ft of uninterrupted brick above lintel and 6 ft of uninterrupted brick on both sides of lintel

"D"—Numbers are only valid if there is 7 ½ ft of uninterrupted brick above lintel and 6 ½ ft of uninterrupted brick on both sides of lintel

"E"—Numbers are only valid if there is 8 ft of uninterrupted brick above lintel and 7 ft of uninterrupted brick on both sides of lintel

"F"—Numbers are only valid if there is 8 ½ ft of uninterrupted brick above lintel and 7 ½ ft of uninterrupted brick on both sides of lintel

NOTES:

NOTES:

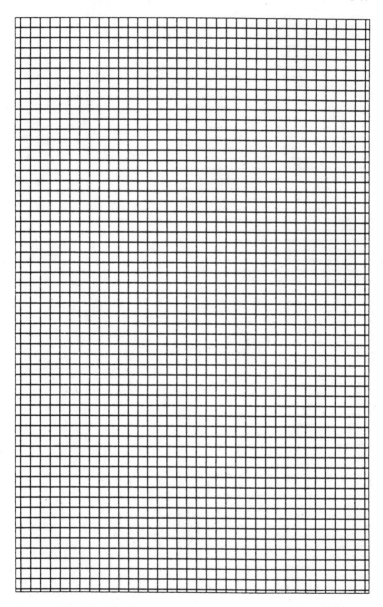

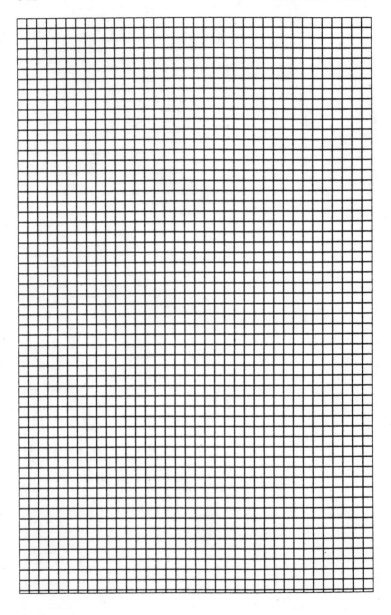

Chapter 9
Geographical Data

TOPICS:
- Seismic
- Wind
- Frost Depth
- Temperature
- Rainfall
- Insulation Requirements
- Insect Susceptibility
- Weathering Potential

TIPS & INFORMATION

- Information is provided in tabular form for quick reference for your particular state.

- Contractors are inundated with advertisements for new literature. This literature often lists sample projects and other testimonials. However, these products may not be as successful in other climates and under differing weather conditions. Refer to the geographical information of the state where successful installations were made to see if climate conditions are similar.

• This section provides information on all 48 contiguous states to assist those in work in various states who are considering expanding into other locales.

	Alabama	Arizona
Seismic Activity	None to minimal	Minimal to considerable
Design Wind Speed (mph)	70–100	70–80
Tornado Frequency	Significant	Minimal
Climate Zone	Cooling	Mixed
Frost Depth (in.)	Negligible	Minimal to 36
Ground Snow Load (psf)	0 to 10	Special
Termites	Very heavy	Moderate to heavy
Decay Susceptibility	Moderate to severe	None to slight
Brick Weathering	Moderate to severe	Negligible to moderate
Concrete Weathering	Negligible to moderate	Negligible to severe
Annual Rainfall (in.)	50 plus	Less than 30"
Time Below 32°	Minimal	Minimal to negligible
Recommended R-Value (Walls)	13–19	13–19
Recommended R-Value (Ceilings)	26	26–33
Radon	All zones	Zone 2
Building Code Zone	Standard	Uniform
Mean Last Date of 32° Temperature	End of March	March
Mean First Date of 32° Temperature	End of October	End of October
Annual Number of Days Above 90° Temperature	90	2 to 150
Number of Days of Precipitation per Year	120	80
% of Possible Time that Sunshine has Occurred	65	80

	Arkansas	California
Seismic Activity	Minimal to considerable	All zones
Design Wind Speed (mph)	70	70–80
Tornado Frequency	Significant	Minimal
Climate Zone	Mixed	Mixed
Frost Depth (in.)	Negligible to 10	Negligible to 37"
Ground Snow Load (psf)	5 to 10	Special
Termites	Moderate to heavy	Moderate to very heavy
Decay Susceptibility	None to slight	None to severe
Brick Weathering	Negligible to moderate	Negligible to severe
Concrete Weathering	Negligible to severe	Negligible to severe
Annual Rainfall (in.)	Less than 30"	60" plus
Time Below 32°	Minimal to negligible	Negligible to minimal
Recommended R-Value (Walls)	13–19	11–19
Recommended R-Value (Ceilings)	26–33	19–33
Radon	Zones 2 and 3	All zones
Building Code Zone	Standard	Uniform
Mean Last Date of 32° Temperature	End of March	Varies
Mean First Date of 32° Temperature	End of October	Varies
Annual Number of Days Above 90° Temperature	75	0 to 170
Number of Days of Precipitation per Year	110	20 to 200
% of Possible Time that Sunshine has Occurred	65	50 to 90

	Colorado	Connecticut
Seismic Activity	None to minimal	Moderate
Design Wind Speed (mph)	70–90	80–90
Tornado Frequency	Mixed to significant	Moderate
Climate Zone	Mixed	Mixed
Frost Depth (in.)	10–60	10–20
Ground Snow Load (psf)	Special	20–35
Termites	None to heavy	Moderate to heavy
Decay Susceptibility	None to slight	Slight to moderate
Brick Weathering	Moderate to severe	Severe
Concrete Weathering	Severe	Severe
Annual Rainfall (in.)	Under 30	60" plus
Time Below 32°	Negligible to significant	Minimal
Recommended R-Value (Walls)	19	19
Recommended R-Value (Ceilings)	30–38	30
Radon	Zones 1 and 2	All zones
Building Code Zone	Uniform	Basic
Mean Last Date of 32° Temperature	June	Early May
Mean First Date of 32° Temperature	August	End of September
Annual Number of Days Above 90° Temperature	0 to 60	10
Number of Days of Precipitation per Year	80 to 150	140
% of Possible Time that Sunshine has Occurred	65	60

	Delaware	Florida
Seismic Activity	Moderate	None
Design Wind Speed (mph)	70–90	90–110
Tornado Frequency	Minimal	Significant
Climate Zone	Heating	Cooling
Frost Depth (in.)	Minimal to 15	Negligible
Ground Snow Load (psf)	15	0
Termites	Moderate to heavy	Very heavy
Decay Susceptibility	Slight to moderate	Moderate to severe
Brick Weathering	Severe	Negligible to moderate
Concrete Weathering	Severe	Negligible
Annual Rainfall (in.)	60" plus	50" plus
Time Below 32°	Minimal	Negligible
Recommended R-Value (Walls)	19	13
Recommended R-Value (Ceilings)	30	26
Radon	All zones	All zones
Building Code Zone	Basic	Standard
Mean Last Date of 32° Temperature	End of April	Mid February
Mean First Date of 32° Temperature	End of September	End of November
Annual Number of Days Above 90° Temperature	10	2 to 10
Number of Rain Days Per Year	140	115
% of Possible Time that Sunshine has Occurred	60	65

	Georgia	Idaho
Seismic Activity	None to moderate	Moderate to considerable
Design Wind Speed (mph)	90–110	70–80
Tornado Frequency	Significant	Minimal
Climate Zone	Cooling	Heating
Frost Depth (in.)	Negligible	6 to 6
Ground Snow Load (psf)	0	Special
Termites	Very heavy	Slight to moderate
Decay Susceptibility	Moderate to severe	None to slight
Brick Weathering	Negligible to moderate	Moderate to severe
Concrete Weathering	Negligible	Severe
Annual Rainfall (in.)	60" plus	Under 30
Time Below 32°	Negligible	Minimal to significant
Recommended R-Value (Walls)	13	18
Recommended R-Value (Ceilings)	26	33–38
Radon	All zones	All zones
Building Code Zone	Standard	Uniform
Mean Last Date of 32° Temperature	End of March	Early June
Mean First Date of 32° Temperature	Mid November	End of August
Annual Number of Days Above 90° Temperature	90	5 to 30
Number of Rain Days Per Year	125	90 to 150
% of Possible Time that Sunshine has Occurred	65	55

	Illinois	Indiana
Seismic Activity	Minimal to Considerable	Minimal to moderate
Design Wind Speed (mph)	70–80	70–80
Tornado Frequency	Significant	Significant
Climate Zone	Heating	Heating
Frost Depth (in.)	6 to 40	6 to 40
Ground Snow Load (psf)	10 to 30	15–30
Termites	Moderate to heavy	Moderate to heavy
Decay Susceptibility	Slight to moderate	Slight to moderate
Brick Weathering	Severe	Severe
Concrete Weathering	Severe	Severe
Annual Rainfall (in.)	30 to 50	30 to 50
Time Below 32°	Minimal to moderate	Minimal to moderate
Recommended R-Value (Walls)	19	19
Recommended R-Value (Ceilings)	30–33	30–33
Radon	Zones 1 and 2	Zones 1 and 2
Building Code Zone	Basic	Uniform
Mean Last Date of 32° Temperature	End of April	End of April
Mean First Date of 32° Temperature	Mid October	Mid October
Annual Number of Days Above 90° Temperature	25	25
Number of Rain Days Per Year	120	130
% of Possible Time that Sunshine has Occurred	60	60

	Iowa	Kansas
Seismic Activity	None to moderate	None to moderate
Design Wind Speed (mph)	80–90	80–90
Tornado Frequency	Significant	Significant
Climate Zone	Heating	Heating
Frost Depth (in.)	25–40	10–30
Ground Snow Load (psf)	20–50	10–25
Termites	Slight to heavy	Moderate to heavy
Decay Susceptibility	None to moderate	None to moderate
Brick Weathering	Moderate to severe	Moderate to severe
Concrete Weathering	Severe	Severe
Annual Rainfall (in.)	30–50	30–50
Time Below 32°	Moderate to significant	Minimal to moderate
Recommended R-Value (Walls)	19	19
Recommended R-Value (Ceilings)	30–33	30–33
Radon	Zone 1	Zones 1 and 2
Building Code Zone	Uniform	Uniform
Mean Last Date of 32° Temperature	Early May	End of April
Mean First Date of 32° Temperature	Early October	Mid October
Annual Number of Days Above 90° Temperature	30	60
Number of Rain Days Per Year	105	90
% of Possible Time that Sunshine has Occurred	60	70

	Kentucky	Louisiana
Seismic Activity	Minimal to considerable	None to minimal
Design Wind Speed (mph)	70–80	70–110
Tornado Frequency	Moderate	Significant
Climate Zone	Cooling	Mixed
Frost Depth (in.)	Negligible to 15	Negligible
Ground Snow Load (psf)	10–15	0–5
Termites	Moderate to heavy	Very heavy
Decay Susceptibility	Slight to moderate	Slight to severe
Brick Weathering	Severe	Moderate to severe
Concrete Weathering	Severe	Moderate to severe
Annual Rainfall (in.)	30 to 50	30 to 60
Time Below 32°	Minimal	Negligible
Recommended R-Value (Walls)	19	13–10
Recommended R-Value (Ceilings)	26–30	26
Radon	All zones	Zone 3
Building Code Zone	Basic	Standard
Mean Last Date of 32° Temperature	End of April	Mid March
Mean First Date of 32° Temperature	End of October	End of November
Annual Number of Days Above 90° Temperature	30	60
Number of Rain Days Per Year	130	105
% of Possible Time that Sunshine has Occurred	60	65

	Maine	Maryland
Seismic Activity	Minimal to moderate	Minimal
Design Wind Speed (mph)	70–100	80–100
Tornado Frequency	Minimal	Minimal
Climate Zone	Heating	Mixed
Frost Depth (in.)	36 to 72	Minimal to 15
Ground Snow Load (psf)	60 to 100	10 to 25
Termites	Moderate to none	Moderate to heavy
Decay Susceptibility	None to slight	Moderate to severe
Brick Weathering	Severe	Severe
Concrete Weathering	Severe	Severe
Annual Rainfall (in.)	60 plus	60 plus
Time Below 32°	Moderate to significant	Minimal
Recommended R-Value (Walls)	19	19
Recommended R-Value (Ceilings)	38	26–30
Radon	Zones 1 and 2	Zones 1 and 2
Building Code Zone	Other	Basic
Mean Last Date of 32° Temperature	End of May	End of April
Mean First Date of 32° Temperature	End of September	End of October
Annual Number of Days Above 90° Temperature	5	10
Number of Rain Days Per Year	160	150
% of Possible Time that Sunshine has Occurred	50	60

	Massachusetts	Michigan
Seismic Activity	Moderate	None
Design Wind Speed (mph)	70–90	80
Tornado Frequency	Significant	Significant
Climate Zone	Heating	Heating
Frost Depth (in.)	20 to 30	25 to 50
Ground Snow Load (psf)	20 to 40	40—70
Termites	Moderate to heavy	None to heavy
Decay Susceptibility	Slight to moderate	None to moderate
Brick Weathering	Severe	Severe
Concrete Weathering	Severe	Severe
Annual Rainfall (in.)	30–50	30–50
Time Below 32°	Moderate	Significant
Recommended R-Value (Walls)	19	19
Recommended R-Value (Ceilings)	33	33–38
Radon	Zones 1 and 2	Zones 1 and 2
Building Code Zone	Basic	Other
Mean Last Date of 32° Temperature	Early May	Early June
Mean First Date of 32° Temperature	End of September	End of September
Annual Number of Days Above 90° Temperature	10	2 to 20
Number of Rain Days Per Year	140	120
% of Possible Time that Sunshine has Occurred	60	50

	Minnesota	Mississippi
Seismic Activity	None	None to minimal
Design Wind Speed (mph)	70–100	70–100
Tornado Frequency	Significant	Significant
Climate Zone	Heating	Cooling
Frost Depth (in.)	40–72	Minimal
Ground Snow Load (psf)	40–70	0–10
Termites	None to slight	Very heavy
Decay Susceptibility	None to moderate	Moderate to severe
Brick Weathering	Severe	Moderate to severe
Concrete Weathering	Severe	Negligible to moderate
Annual Rainfall (in.)	30 to 50	60 and over
Time Below 32°	Significant	Negligible
Recommended R-Value (Walls)	19	13–19
Recommended R-Value (Ceilings)	33–38	26
Radon	Zones 1 and 2	Zones 2 and 3
Building Code Zone	Uniform	Standard
Mean Last Date of 32° Temperature	End of May	Mid March
Mean First Date of 32° Temperature	Mid September	Early November
Annual Number of Days Above 90° Temperature	10	90
Number of Rain Days Per Year	110	110
% of Possible Time that Sunshine has Occurred	60	60

	Missouri	Montana
Seismic Activity	Minimal to considerable	Minimal
Design Wind Speed (mph)	70–90	70–90
Tornado Frequency	Significant	Minimal
Climate Zone	Heating	Heating
Frost Depth (in.)	6 to 30	13 to 60
Ground Snow Load (psf)	10 to 25	Special
Termites	Moderate to heavy	None to slight
Decay Susceptibility	Slight to moderate	None to slight
Brick Weathering	Severe	Moderate to severe
Concrete Weathering	Severe	Severe
Annual Rainfall (in.)	30 to 50	Less than 30
Time Below 32°	Moderate	Significant
Recommended R-Value (Walls)	19	19
Recommended R-Value (Ceilings)	26 to 30	33–38
Radon	Zones 1 and 2	Zones 2 and 3
Building Code Zone	Basic	Uniform
Mean Last Date of 32° Temperature	Mid April	End of May
Mean First Date of 32° Temperature	End of October	Mid September
Annual Number of Days Above 90° Temperature	60	2 to 30
Number of Rain Days Per Year	110	90 to 120
% of Possible Time that Sunshine has Occurred	65	60

	Nebraska	Nevada
Seismic Activity	Moderate	Moderate to high
Design Wind Speed (mph)	80	70–90
Tornado Frequency	Significant	Minimal
Climate Zone	Heating	Heating
Frost Depth (in.)	15–40	Minimal to 22
Ground Snow Load (psf)	10–25	Special
Termites	Slight to heavy	Slight to heavy
Decay Susceptibility	Slight to heavy	Slight to heavy
Brick Weathering	Moderate to severe	Moderate to severe
Concrete Weathering	Severe	Moderate to severe
Annual Rainfall (in.)	Up to 50	Less than 30
Time Below 32°	Moderate	Minimal to moderate
Recommended R-Value (Walls)	19	19
Recommended R-Value (Ceilings)	30–33	26–33
Radon	Zones 1 and 2	All zones
Building Code Zone		
Mean Last Date of 32° Temperature	Mid May	End of May
Mean First Date of 32° Temperature	End of September	End of August
Annual Number of Days Above 90° Temperature	45	5 to 60
Number of Rain Days Per Year	90	30 to 90
% of Possible Time that Sunshine has Occurred	65	80

	New Hampshire	New Jersey
Seismic Activity	Moderate	Minimal
Design Wind Speed (mph)	70–90	70–90
Tornado Frequency	Minimal	Moderate
Climate Zone	Heating	Heating
Frost Depth (in.)	48–60	10–25
Ground Snow Load (psf)	40–60	20–40
Termites	None to heavy	Moderate to heavy
Decay Susceptibility	None to slight	Slight to moderate
Brick Weathering	Severe	Severe
Concrete Weathering	Severe	Severe
Annual Rainfall (in.)	30 to 50	30 to 50
Time Below 32°	Significant	Moderate
Recommended R-Value (Walls)	19	19
Recommended R-Value (Ceilings)	33–38	30
Radon	Zone 1	All zones
Building Code Zone		
Mean Last Date of 32° Temperature	End of April	End of April
Mean First Date of 32° Temperature	Mid September	Mid September
Annual Number of Days Above 90° Temperature	10	10
Number of Rain Days Per Year	140	140
% of Possible Time that Sunshine has Occurred	60	60

	New Mexico	New York
Seismic Activity		
Design Wind Speed (mph)	70–90	70–90
Tornado Frequency	Minimal	Moderate
Climate Zone	Mixed	Heating
Frost Depth (in.)	Minimal to 24	15–60
Ground Snow Load (psf)	Special	30–50
Termites	Moderate to heavy	None to heavy
Decay Susceptibility	None to slight	None to moderate
Brick Weathering	Negligible to moderate	Severe
Concrete Weathering	Negligible to severe	Severe
Annual Rainfall (in.)	Less than 30	30 to 50
Time Below 32°	Minimal	Moderate
Recommended R-Value (Walls)	19	19
Recommended R-Value (Ceilings)	26–38	30–38
Radon	Zones 1 and 2	All zones
Building Code Zone		Other
Mean Last Date of 32° Temperature	End of May	Mid May
Mean First Date of 32° Temperature	Early October	End of September
Annual Number of Days Above 90° Temperature	2 to 90	10
Number of Rain Days Per Year	80	140 to 180
% of Possible Time that Sunshine has Occurred	80	60

	North Carolina	North Dakota
Seismic Activity	Minimal to moderate	None
Design Wind Speed (mph)	70–110	80–100
Tornado Frequency	Moderate	Moderate
Climate Zone	Mixed	Heating
Frost Depth (in.)	0 to 15	35–60
Ground Snow Load (psf)	10–20	25–50
Termites	Moderate to heavy	None to moderate
Decay Susceptibility	Slight to severe	None to slight
Brick Weathering	Moderate to severe	Moderate to severe
Concrete Weathering	Moderate to severe	Severe
Annual Rainfall (in.)	30 to over 60	As much as 50
Time Below 32°	Minimal	Moderate to significant
Recommended R-Value (Walls)	19	19
Recommended R-Value (Ceilings)	26–30	38
Radon	All zones	Zone 1
Building Code Zone		
Mean Last Date of 32° Temperature	Early April	End of May
Mean First Date of 32° Temperature	Early November	Mid September
Annual Number of Days Above 90° Temperature	5 to 30	20
Number of Rain Days Per Year	120	100
% of Possible Time that Sunshine has Occurred	65	60

	Ohio	Oklahoma
Seismic Activity	Minimal	None to moderate
Design Wind Speed (mph)	70–80	70–90
Tornado Frequency	Moderate	Extreme
Climate Zone	Heating	Mixed
Frost Depth (in.)	10–30	Minimal to 15
Ground Snow Load (psf)	Special	5–20
Termites	Moderate to heavy	Moderate to heavy
Decay Susceptibility	Slight to moderate	None to moderate
Brick Weathering	Severe	Moderate to severe
Concrete Weathering	Severe	Moderate to severe
Annual Rainfall (in.)	30–50	30–50
Time Below 32°	Moderate	Minimal
Recommended R-Value (Walls)	19	19
Recommended R-Value (Ceilings)	30–33	26–30
Radon	Zones 1 and 2	Zones 2 and 3
Building Code Zone		
Mean Last Date of 32° Temperature	Early May	Mid April
Mean First Date of 32° Temperature	Mid October	End of October
Annual Number of Days Above 90° Temperature	20	90
Number of Rain Days Per Year	140	90
% of Possible Time that Sunshine has Occurred	50	70

	Oregon	Pennsylvania
Seismic Activity	Moderate to considerable	Minimal to moderate
Design Wind Speed (mph)	70–90	70–80
Tornado Frequency	Minimal	Moderate
Climate Zone	Heating	Heating
Frost Depth (in.)	None to 21	15–40
Ground Snow Load (psf)	Special	Special
Termites	Slight to heavy	Moderate to heavy
Decay Susceptibility	None to moderate	Slight to moderate
Brick Weathering	Moderate to severe	Severe
Concrete Weathering	Moderate to severe	Severe
Annual Rainfall (in.)	Under 30 to over 60	30–50
Time Below 32°	Negligible to moderate	Minimal to moderate
Recommended R-Value (Walls)	19	19
Recommended R-Value (Ceilings)	19–33	30–33
Radon	All zones	Zones 2 and 3
Building Code Zone		
Mean Last Date of 32° Temperature	End of June	End of May
Mean First Date of 32° Temperature	Early August	End of September
Annual Number of Days Above 90° Temperature	2 to 30	5 to 30
Number of Rain Days Per Year	90 to 180	170
% of Possible Time that Sunshine has Occurred	60	60

	Rhode Island	South Carolina
Seismic Activity	Moderate	Moderate
Design Wind Speed (mph)	80–90	70–110
Tornado Frequency	Moderate	Moderate
Climate Zone	Heating	Cooling
Frost Depth (in.)	10–20	Less than 6 in.
Ground Snow Load (psf)	20–35	0–15
Termites	Moderate to heavy	Very heavy
Decay Susceptibility	Slight to moderate	Slight to severe
Brick Weathering	Severe	Moderate to severe
Concrete Weathering	Severe	Moderate
Annual Rainfall (in.)	More than 60	30 to over 60
Time Below 32°	Minimal to moderate	Negligible
Recommended R-Value (Walls)	19	19
Recommended R-Value (Ceilings)	30	26–30
Radon	Zones 1 and 2	All zones
Building Code Zone		
Mean Last Date of 32° Temperature	End of May	End of March
Mean First Date of 32° Temperature	Early October	Mid November
Annual Number of Days Above 90° Temperature	10	30
Number of Rain Days Per Year	140	110
% of Possible Time that Sunshine has Occurred	60	

	South Dakota	Tennessee
Seismic Activity	None to minimal	Minimal to considerable
Design Wind Speed (mph)	80–100	70
Tornado Frequency	Significant	Moderate
Climate Zone	Heating	Mixed
Frost Depth (in.)	20–60	0 to 15
Ground Snow Load (psf)	15–50	Special
Termites	Slight to heavy	Moderate to heavy
Decay Susceptibility	None to slight	Slight to severe
Brick Weathering	Moderate	Severe
Concrete Weathering	Severe	Moderate to severe
Annual Rainfall (in.)	Under 30 up to 50	50–60
Time Below 32°	Moderate to significant	Minimal
Recommended R-Value (Walls)	19	19
Recommended R-Value (Ceilings)	33–38	26
Radon	Zones 1 and 2	All zones
Building Code Zone		
Mean Last Date of 32° Temperature	Mid May	End of April
Mean First Date of 32° Temperature	End of September	End of October
Annual Number of Days Above 90° Temperature	30	60
Number of Rain Days Per Year	100	120
% of Possible Time that Sunshine has Occurred	60	60

	Texas	Utah
Seismic Activity	None to minimal	Minimal to considerable
Design Wind Speed (mph)	70–100	70–80
Tornado Frequency	Extreme	Minimal
Climate Zone	Cooling/mixed	Heating
Frost Depth (in.)	Minimal to 10	Up to 24
Ground Snow Load (psf)	0 to 20	Special
Termites	Moderate to very heavy	Slight to heavy
Decay Susceptibility	None to moderate	None to slight
Brick Weathering	Negligible to moderate	Moderate to severe
Concrete Weathering	Negligible to severe	Severe
Annual Rainfall (in.)	0 to 50	Under 30
Time Below 32°	Negligible	Moderate
Recommended R-Value (Walls)	13–19	19
Recommended R-Value (Ceilings)	26–30	30–33
Radon	Zones 2 and 3	Zones 1 and 2
Building Code Zone		
Mean Last Date of 32° Temperature	Early April	End of June
Mean First Date of 32° Temperature	End of November	End of August
Annual Number of Days Above 90° Temperature	90 to 150	5 to 90
Number of Rain Days Per Year	60 to 100	90
% of Possible Time that Sunshine has Occurred	70	70

	Vermont	Virginia
Seismic Activity	Moderate	Minimal to moderate
Design Wind Speed (mph)	70–80	70–100
Tornado Frequency	Minimal	Moderate
Climate Zone	Heating	Mixed
Frost Depth (in.)	Up to 60	Up to 15
Ground Snow Load (psf)	Up to 60	Special
Termites	None to moderate	Moderate to heavy
Decay Susceptibility	None to moderate	Moderate to severe
Brick Weathering	Severe	Moderate to severe
Concrete Weathering	Severe	Moderate to severe
Annual Rainfall (in.)	50 to 60	Up to 60
Time Below 32°	Moderate	Minimal
Recommended R-Value (Walls)	19	19
Recommended R-Value (Ceilings)	33–38	26–30
Radon	Zones 2 and 3	All zones
Building Code Zone		
Mean Last Date of 32° Temperature	Mid May	End of April
Mean First Date of 32° Temperature	Mid September	Mid October
Annual Number of Days Above 90° Temperature	10	50
Number of Rain Days Per Year	150	140
% of Possible Time that Sunshine has Occurred	60	60

	Washington	West Virginia
Seismic Activity	Moderate to risk	Minimal
Design Wind Speed (mph)	70 to special	70
Tornado Frequency	Minimal	Minimal
Climate Zone	Heating	Mixed
Frost Depth (in.)	Up to 21	Up to 20
Ground Snow Load (psf)	Special	Special
Termites	None to moderate	Moderate to heavy
Decay Susceptibility	None to moderate	Slight to moderate
Brick Weathering	Moderate to severe	Severe
Concrete Weathering	Moderate to severe	Severe
Annual Rainfall (in.)	Under 30 and over 60	30 to 60
Time Below 32°	Minimal to moderate	Minimal
Recommended R-Value (Walls)	19	19
Recommended R-Value (Ceilings)	19–33	30
Radon	All zones	All zones
Building Code Zone		
Mean Last Date of 32° Temperature	End of May	End of March
Mean First Date of 32° Temperature	Early October	Mid October
Annual Number of Days Above 90° Temperature	0 to 30	50
Number of Rain Days Per Year	90 to 200	140
% of Possible Time that Sunshine has Occurred	50	65

	Wisconsin	Wyoming
Seismic Activity	None	None to considerable
Design Wind Speed (mph)	70 to 90	Special
Tornado Frequency	Minimal to significant	Minimal
Climate Zone	Heating	Heating
Frost Depth (in.)	Up to 54	Up to 54
Ground Snow Load (psf)	60 to special	Special
Termites	Slight to moderate	None to slight
Decay Susceptibility	Slight to moderate	None to slight
Brick Weathering	Severe	Moderate
Concrete Weathering	Severe	Severe
Annual Rainfall (in.)	50 to 60	Under 30
Time Below 32°	Significant	Moderate to significant
Recommended R-Value (Walls)	19	19
Recommended R-Value (Ceilings)	33–38	33–38
Radon	Zones 1 and 3	Zones 1 and 2
Building Code Zone		
Mean Last Date of 32° Temperature	End of May	End of June
Mean First Date of 32° Temperature	Mid September	End of August
Annual Number of Days Above 90° Temperature	20	5 to 20
Number of Rain Days Per Year	120	90 to 150
% of Possible Time that Sunshine has Occurred	50	60

Chapter 10
Miscellaneous

TIPS & INFORMATION

- The design drawings should clearly show the details of the roof framing. Letting the carpenter contractor decide how to do the framing can lead to long term problems.

- Attic spaces need to be properly vented to allow the hot air to be removed during the summer. Venting should include venting near the roof line as well as at the soffit.

- Be sure to properly nail roof members together. Building codes identify the number and sizes of nails needed.

- Note that a ridge beam and a ridge board are not the same member. A ridge beam must transfer the load from the rafters to an underlying support. A ridge board is only used to nail the ends of the rafters and is not designed to transfer loads.

- In areas of high winds, the roof rafters may need to be fastened to the walls with straps that resist uplift.

- Septic systems should be designed by a properly trained person. Understanding the permeability of the soil is a must to design a system that disperses properly.

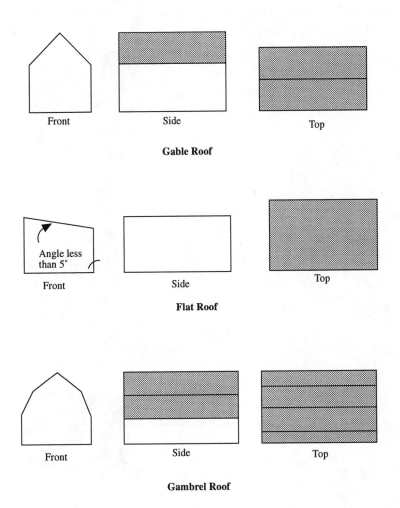

Figure 10.1—Types of Roof Structures

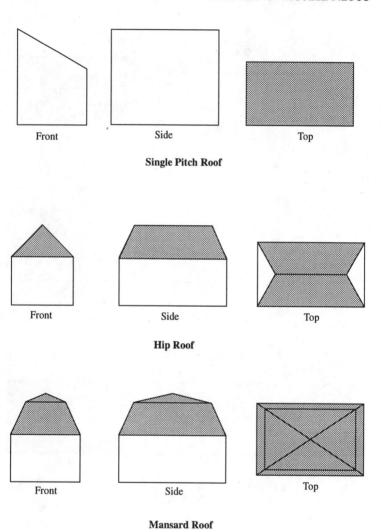

Figure 10.1—Types of Roof Structures continued

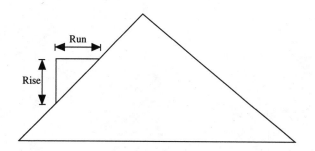

Slope	Rise	Run	Angle (degrees)
1 in 12	1	12	5
1 in 6	2	12	9
1 in 4	3	12	14
1 in 3	4	12	18
5 in 12	5	12	23
1 in 12	6	12	27
7 in 12	7	12	30
2 in 3	8	12	34
3 in 4	9	12	37
5 in 6	10	12	40
11 in 12	11	12	43
1 in 1	12	12	45

Roof Slopes

Top Half

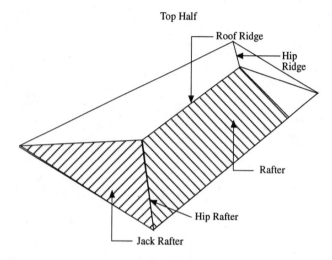

Hip Roof Framing

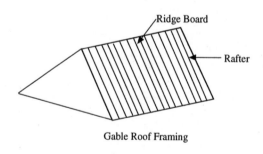

Gable Roof Framing

Bottom Half

Roof Framing Nomenclature

Roofing Materials

Type	Material	Description	Life (years)
Sloped	Asphalt	Organic material saturated in asphalt and covered with granules	10–30
Sloped	Slate	Stone material made into shingles	50–100
Sloped	Wood	Cedar or redwood that is split into shingles	Varies
Sloped	Tile	Clay or concrete made into flat pieces or barred shapes	50+
Sloped	Metal	Corrugated or flat sheets of metal	Varies
Flat	EPDM	Rubber sheet that is monolithic roof cover	Varies
Flat	Modified Bitumen	Roll of asphalt based material	Varies

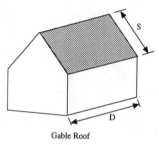

Gable Roof

Roof area = $2 \times S \times D$

Squares needed with 10% Waste
= $0.022 \times S \times D$

- For 50 ft deep building with
  rake length of 15 ft requires:

    = $0.022 \times 15 \times 50$

    = 16.5 use 17 squares

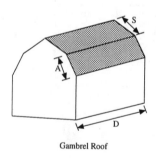

Gambrel Roof

Roof area = $2(AD + SD)$

Squares needed with 10% Waste
= $0.022 \times D \times (A + S)$

- For 50 ft deep building with
  lower rake of 12 ft and upper
  rake of 10 ft requires:

    = $0.022 \times 50 \times (10 + 12)$

    = 24.2 use 24 squares

Figure 10.3—Roofing Material Determination

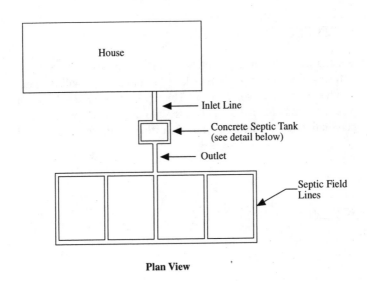

Plan View

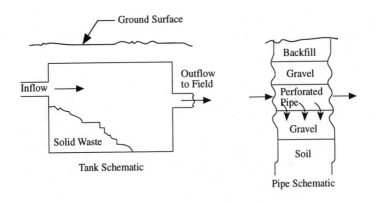

Figure 10.4—Septic System Layout

SOIL PERCOLATION TEST

PURPOSE:

Wastewater will flow from the house into the septic tank. Solids will sink to the bottom of the tank and liquid effluent will flow out into the pipes of the septic field. The liquid in the pipe will exit the pipe through holes and then disperse into the soil below. The percolation test predicts how fast the effluent will flow into the soil.

SET UP:

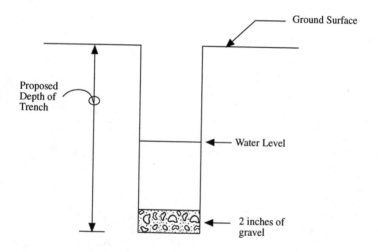

CONCEPT:

A test hole is dug in the area where the septic field will be installed. The hole is saturated for a certain time prior to starting the test. Water is then placed in the hole and its rate of disbursement is determined.

PRE-TEST SATURATION:

Add 12 inches of water 24 hours before test will commence. If water seeps out, refill water level.

TEST:

Adjust water depth to 6 inches. Run test for 3 1/2 hours by continuing to bring water level back to 6 inches. Record water drop between 3 1/2 and 4 hours. This value is the rate that will be utilized.

RESULTS:

The required absorption area per bedroom can be roughly estimated as

$$\text{Area} = \frac{120}{R} + 125$$

where R equals the amount the water dropped in last half hour given in inches.

NOTES:

The information provided is for a rough guideline only. A properly functioning septic system should be designed by a trained professional.

Square Roots, Squares and Cubes

Number "X"	Square Root ($\sqrt{X}$)	Square (X times X)	Cube (X times X times X)
1	1	1	1
2	1.4	4	8
3	1.73	9	27
4	2	16	64
5	2.24	25	125
6	2.45	36	216
7	2.65	49	343
8	2.83	64	512
9	3	81	729
10	3.16	100	1000
11	3.32	121	1331
12	3.46	144	1728
13	3.61	169	2197
14	3.74	196	2477
15	3.87	225	3375
16	4	256	4096
17	4.12	289	4913
18	4.24	324	5832
19	4.36	361	6859
20	4.47	400	8000
21	4.58	441	9621
22	4.69	484	10648
23	4.80	529	12167
24	4.90	576	13824
25	5	625	15625

Properties of Circles

Diameter	Radius	Area	Perimeter
1.0	0.5	0.79	3.14
1.5	0.75	1.77	4.71
2.0	1.0	3.14	6.28
2.5	1.25	4.91	7.85
3.0	1.5	7.07	9.42
3.5	1.75	9.62	10.99
4.0	2.0	12.57	12.56
4.5	2.25	15.90	14.13
5.0	2.25	19.64	15.7
5.5	2.75	23.76	17.27
6.0	3.0	28.27	18.84
6.5	3.25	33.18	20.42
7.0	3.5	38.48	21.99
7.5	3.75	44.18	23.56
8.0	4.0	50.27	25.13
8.5	4.25	56.75	26.70
9.0	4.5	63.62	28.27
9.5	4.75	70.88	29.84
10.0	5.0	78.54	31.41

Temperature Conversions

Degrees Celsius	Degrees Fahrenheit	Degrees Celsius	Degrees Fahrenheit
-40	-40	35	95
-35	-31	40	104
-30	-22	45	113
-25	-13	50	122
-20	-4	55	131
-15	5	60	140
-10	14	65	149
-5	23	70	158
0	32	75	167
5	41	80	176
10	50	85	185
15	59	90	194
20	68	95	203
25	77	100	212
30	86	105	221

Fraction Conversions

Fraction	Decimal	Fraction	Decimal
1/64	0.016	1/4	0.25
1/32	0.031	17/64	0.266
3/64	0.047	9/32	0.281
1/16	0.063	19/64	0.297
5/64	0.078	5/16	0.313
3/32	0.094	21/64	0.328
7/64	0.109	11/32	0.344
1/8	0.125	23/64	0.359
9/64	0.141	3/8	0.375
5/32	0.156	25/64	0.391
11/64	0.172	13/32	0.406
3/16	0.188	27/64	0.421
13/64	0.203	7/16	0.438
7/32	0.219	29/64	0.453
15/64	0.234	15/32	0.469

Fraction Conversion

Fraction	Decimal	Fraction	Decimal
31/64	0.484	3/4	0.75
1/2	.5	49/64	0.766
33/64	0.516	25/32	0.781
17/32	0.531	51/64	0.797
35/64	0.547	13/16	0.813
9/32	0.281	53/64	0.828
37/64	0.578	27/32	0.844
19/32	0.594	55/64	0.859
39/64	0.609	7/8	0.875
5/8	0.625	57/64	0.891
41/64	0.641	29/32	0.906
21/32	0.656	59/64	0.922
43/64	0.672	15/16	0.938
11/16	0.688	61/64	0.953
45/64	0.703	31/32	0.969
23/32	0.719	63/64	0.984
47/64	0.734	1.0	1.0

Common Abbreviations

Average	AVE
Amer. Concrete Institute	ACI
Amer. Inst. of Steel Construction	AISC
Amer. Inst. of Timber Construction	AITC
Amer. Plywood Association	APA
Amer. Society of Civil Engineers	ASCE
American Welding Society	AWS
Assoc. General Contractors of Amer.	AGC
Brick Institute of America	BIA
Clay Pipe	CP
Construction Specification Institute	CSI
Construction Joint	CJ
Dead Load	DL
Degree	DEG
Diameter	DIA
Dimension	Dim
Dishwasher	DW
Double Hung Window	DH
Downspout	DS
Figure	FIG
Flange	FLG
Floor Drain	FD
Forest Products Laboratory	FPL
Foundation	FDN
Gauge	GA
Hardware	HDW
Header	HDR
Height	HGT
Highway	HWY

Common Abbreviations (cont.)

Inside Diameter	ID
Joint	JT
Lavatory	LAV
Live Load	LL
Membrane	MEM
Mirror	MIR
Miscellaneous	MISC
Nat. Assoc. of Home Builders	NHB
National Bureau of Standards	NBS
National Forest Products Assoc.	NFPA
National Fire Products Assoc.	NFPA
National Research Council	NRC
National Roofing Contr. Assoc.	NRCA
National Electric Code	NEC
Nonslip Tread	NST
Occupational Safety & Health Adm.	OSHA
Not To Scale	NTS
Opening	OPNG
Opposite	OPP
Outside Diameter	OD
Overhead	OH
Penny (nail)	d
Piling	PLG
Plumbing	PLMB
Plywood	PLYWD
Portland Cement	PC
Portland Cement Association	PCA
Precast	PRCT
Precast Concrete Institute	PCI

Common Abbreviations (cont.)

Property Line	PL
Quality	QUAL
Pump	PMP
Radius	R
Reinforce	REIF
Reinforced Concrete	RC
Relative Humidity	RH
Right-of-way	ROW
Roofing	RFG
Safety	SAF
Sanitary	SAN
Sewer	SEW
Shower	SH
Siding	SDG
Sing	SK
Soundproof	SNDPRF
Stairway	STWY
Standard	STD
Steel Joist Institute	SJI
Strength	STR
Telephone	TEL
Toilet	T
Tongue and Groove	T & G
Total Load	TL
Typical	TYP
Uniform Building Code	UBC
Underwriters Laboratory	UL

Common Units

Acre	acr
Amp	amp
Celsius	c
Centimeter	cm
Day	d
Decibel	dB
Fahrenheit	f
Feet	ft
Gallon	gal
Hertz	hz
Horsepower	hp
Hour	hr
Inch	in
Meters	m
Millimeter	mm
Miles	mi
Minute	min
Month	mo
Ounce	oz
Pint	pt
Pounds	lb
Quart	qt
Second	sec
Square Inch	sq in
Square Foot	sq ft
Square Mile	sq mi
Ton	ton
Watts	w
Week	wk
Yards	yd
Year	yr

Inches to Centimeters

Inches	Centimeters	Inches	Centimeters
1/2	1.27	13	33.02
1	2.54	13-1/2	34.29
1-1/2	3.81	14	35.56
2	5.08	14-1/2	36.83
2-1/2	6.35	15	38.10
3	7.62	15-1/2	39.37
3-1/2	8.89	16	40.64
4	10.16	16-1/2	41.91
4-1/2	11.43	17	43.18
5	12.70	17-1/2	44.45
5-1/2	13.97	18	45.72
6	15.24	18-1/2	46.99
6-1/2	16.51	19	48.26
7	17.78	19-1/2	49.53
7-1/2	19.05	20	50.80
8	20.32	20-1/2	52.07
8-1/2	21.59	21	53.34
9	22.86	21-1/2	54.61
9-1/2	24.13	22	55.88
10	25.4	22-1/2	57.15
10-1/2	26.67	23	58.42
11	27.94	23-1/2	59.69
11-1/2	29.21	24	60.96
12	30.48	24-1/2	62.23
12-1/2	31.75	25	63.50

Centimeters to Inches

Inches	Centimeters	Inches	Centimeters
1	0.039	26	10.236
2	0.079	27	10.630
3	1.181	28	11.024
4	1.575	29	11.417
5	1.969	30	11.811
6	2.362	31	12.205
7	2.756	32	12.598
8	3.150	33	12.992
9	3.543	34	13.386
10	3.937	35	13.780
11	4.331	36	14.173
12	4.724	37	14.567
13	5.118	38	14.961
14	5.512	39	15.354
15	5.906	40	15.748
16	6.299	41	16.142
17	6.693	42	16.535
18	7.087	43	16.929
19	7.480	44	17.323
20	7.874	45	17.717
21	8.268	46	18.110
22	8.661	47	18.504
23	9.055	48	18.898
24	9.449	49	19.291
25	9.843	50	19.685

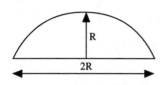

$$A = \frac{\pi R^2}{2}$$

$$I = \frac{\pi R^4}{8}$$

$$Y = \frac{4R}{3\pi}$$

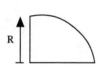

$$A = \frac{\pi R^2}{4}$$

$$I = \frac{\pi R^4}{4}$$

$$Y = \frac{4R}{3\pi}$$

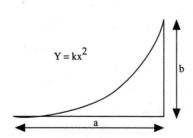

$$A = \frac{ab}{3}$$

$$I = \frac{ab^3}{21}$$

$$Y = \frac{3b}{10}$$

Rectangle

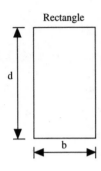

$$A = bd$$

$$I_x = \frac{bd^3}{12}$$

$$I_y = \frac{db^3}{12}$$

$$\overline{Y} = \quad center$$

Triangle

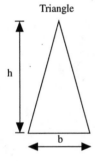

$$A = \frac{bh}{2}$$

$$I = \frac{bh^3}{36}$$

$$\overline{Y} = \frac{h}{3}$$

Circle

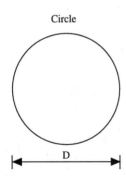

$$A = \frac{\pi D^2}{4}$$

$$I = \frac{\pi D4}{64}$$

$$\overline{Y} = \quad center$$

Chapter 11
Resources (Trade Organizations)

TIPS & INFORMATION

- Trade organizations are typically started and funded by the industry they are dedicated to. They can be a wealth of information, but always remember that they exist because of the product they represent. Therefore, you may not always get a critical view of a product. Nevertheless, they are a valuable source of information.

- Trade organizations often have excellent technical literature at reasonable prices. You may also be able to get reproductions of out of print books.

- Request to speak with a technical person on a particular subject rather than the representatives that handle the inquiry telephone calls.

- Trade organizations often subsidize seminars on the products they represent. Request a seminar schedule or ask if you can receive a copy of seminar handouts.

- Request a list of manufacturers on the product you are interested in. This will give you contacts for larger suppliers who will be more likely to provide additional technical assistance.

- Trade organizations are involved in following and sometimes influencing building code requirements. Although they may be unwilling to interpret a specific code requirement, they can provide information relating to the interpretations of building codes.

- Trade organizations can provide historical background on building code issues which can assist in determining the applicable code at a specific time.

- It is the best interest of trade organizations to assist you in successfully marketing their product. In some cases, stock presentations can be obtained at minimal cost to assist you in marketing clients.

ASPHALT

Asphalt Institute
Research Park Drive
P.O. Box 14052
Lexington, KY 40512
(606) 288-4960

Members consist of refiners of
asphalt products. Publishes
engineering manual and
specifications on the use of
asphalt in construction.

Notes: _____

_____

**Asphalt Roofing
Manufacturers Assoc.**
6000 Executive Drive, Suite 201
Rockville, MD 20852
(301) 231-9050

Members consist of asphalt
shingle manufacturers and
related industries. Provides
information on built-up roofing,
modified bituminous roofing
and residential roofing.

Notes _____

_____

Asphalt Recycling Assoc.
3 Church Circle
Annapolis, MD 21401
(410) 267-0023

Provides information on
recycling of roadway pave-
ments.

Notes: _____

_____

**National Asphalt Pavement
Assoc.**
5100 Forbes Blvd.
Lanham, MD
(301) 731-4748

Provides information on
construction practices and
materials for asphalt used in
highway paving.

Notes: _____

_____

BRICK

Brick Institute of America
11490 Commerce Park Drive
Resont, VA 22091
(703) 620-0010

Provides information on the use
of clay bricks. Extensive
publications on proper construction techniques and design
details.

Notes: _____

_____

**National Association of Brick
Distributors**
1600 Spring Hill Road
Vienna, VA 22182
(703) 729-6223
Fax (703) 729-6227

Members consist of distributors
of clay bricks. Provides
publications on care of brick and
technical information.

Notes: _____

_____

**National Concrete Masonry
Assoc.**
2301 Horse Pen Road
Herndon, VA 22071
(703) 713-1900
www.ncma.org

Provides information on
concrete blocks and paving
blocks. Technical literature for
design of concrete walls and
safety issues.

Notes: _____

_____

CONCRETE

Concrete Sawing and Drilling Assoc.
4900 Blazer Parkway
Dublin, OH 43107
(614) 766-3656

Promotes the use of cutting concrete with diamond tools. Members consist of contractors and affiliated industries.

Notes: _____

_____

Portland Cement Assoc.
5420 Old Orchard Rd.
Skokie, IL 60077
(847) 966-6200

Members consist of portland cement manufacturers and affiliated industries. Provides seminars and literature on all aspects of concrete including residential uses.

Notes: _____

_____

International Concrete Repair Institute
1323 Shepard Drive
Sterling, VA 20164
(703) 450-0116

Provides technical information and seminars for the repair of concrete damaged by a variety of causes. Members consist of contractors, manufacturers and technical professionals.

Notes: _____

_____

Precast Concrete Institute
175 West Jackson
Chicago, IL 60604
(312) 786-0300

Provides information on design and construction of prestressed concrete. Data on standard concrete member sections.

Notes: _____

_____

FLOORING

Ceramic Tile Distributors
800 Roosevelt Rd.
Glen Ellyn, IL 60137
(800) 938-CTDA
Fax (630) 790-3095
www.ctdahome.org

Provides information about floor
tile and installation.

Notes: _____

_____

Maple Flooring Assoc.
60 Revere Dr., Suite 500
Northbrook, IL 60062

Members consist of manufacturers and suppliers of maple
flooring. Sets grading standards
for maple flooring.

Notes: _____

_____

Italian Tile Center
499 Park Avenue
New York, New York, 10022
(212) 980-1500

Provides information for the
buyer of ceramic tiles manufac-
tured in Italy.

Notes: _____

_____

**Oak Flooring Manufacturers
Assoc.**
P.O. Box 3009
Memphis, TN 38173
(901) 526-5016
www.nofma.org

Members consist of manufacturers of oak flooring. Sets grading
standards for oak flooring.

Notes: _____

_____

LUMBER

National Forest Products Assoc.
1250 Connecticut Ave., NW,
Suite 200
Washington, D.C. 20036
(204) 463-2799

Publishes technical data for the
construction of wood structures,
including design specifications
for use by architects and
engineers.

Notes: _____

_____

Southern Forest Products Assoc.
P.O. Box 641700
Kenner, LA 70064
(504) 443-6612
Fax (504) 443-6612
www.sfpa.org

A widely used type of wood is the
southern pine. Members of this
organization consist of manufacturers of southern pine lumber.
Association provides information
on the use and marketing of
southern pine lumber.

Notes: _____

_____

Pacific Lumber Inspection
P.O. Box 7235
Belleue, WA 98008
(206) 746-6542

Service approved by several
standard committees for grading
construction lumber for American and Canadian lumber.

Notes: _____

_____

Southern Pine Inspection Bureau
4709 Scenic Highway
Pensacola, FL 32504
(904) 434-2611
Fax (904) 433-5594

This bureau is sponsored by
southern pine manufacturers and
develops grading standards for
southern pine lumber.

Notes: _____

_____

LUMBER—SPECIALTY

Southern Pressure Treaters Assoc.
P.O. Box 2389
Gulf Shores, AL 36547
(205) 968-5726
Fax (205) 968-6008

The pressure treating of lumber products protects them from decay. Pressure treating is used for utility poles, framing lumber, fence posts, etc. Members of this organization consist of manufacturers of who apply the pressure treatment. Association provides information on product mix plant operations and standards.

Notes: _____

_____

Truss Plate Institute
583 D'Onoforio Dr., Suite 200
Madison, WI 53719
(608) 833-5900

Consists of suppliers of trusses and material for trusses. Provides information on temporary support of trusses during erection and design specifications.

Notes: _____

_____

Wood Truss Council of America
5937 Meadowod Dr., Suite 14
Madison, WI 53711
(608) 274-3329
www.woodtruss.com

Wood trusses are an efficient, reliable, cost effective method for use as roof framing. Members consist of manufacturers and suppliers of trusses and truss components. Council provides technical literature on the use of metal plate connected wood trusses.

Notes: _____

_____

MATERIALS AND TESTING

American Society of Nonde-structive Testing
1711 Arlingate Lane
P.O. Box 28518
Columbus, OH 43228
(614) 274-6003
www.asnt.org

Provides information and education on the testing of materials without damaging the materials. Information provided on techniques for ultrasound, magnetic and other types of testing to identify potential flaws.

Notes: _____

_____

American Society for Testing and Materials
100 Bar Harbor Dr.
West Conshohocken, PA 19428
www.astm.org

Members consist of technical individuals, consumers and technicians. Develops standard-ized test methods for all types of construction materials.

Notes: _____

_____

ASM International
9639 Kinsman
Materials Park, OH 44073
(216) 338-5151

Members consist of technical professionals involved with engineered materials, particu-larly metals. Provides informa-tion on nondestructive testing.

Notes: _____

_____

STEEL

Metal Building Assoc.
1300 Summer Ave.
Cleveland, OH 44115
(216) 241-7333
www.mbma.com

Consists of manufacturers of
pre-engineered metal buildings.
Provides design manual and
specifications for analysis and
design.

Notes: _____

_____

Steel Deck Institute
P.O. 9506
Canton, OH 44711
(216) 493-7886
www.sdi.org

Membership composed of steel
deck manufacturers. Provides
specifications for steel floor and
roof decks including those
composite with concrete.

Notes: _____

_____

Metal Construction Assoc.
1767 Business Center Dr.
Suite 302
Reston, VA 22090
(703) 438-8285
www.mcal.org

Members consist of a wide
variety of individuals engaged in
the metal construction industry.
Provides information on structural erection and estimating.

Notes: _____

_____

Steel Joist Institute
1205 48th Ave. N., Suite A
Myrtle Beach, SC 29577
(803) 449-0487
www.steeljoist.prg

Members consist of manufacturers that produce open web steel
joists and joist girders. Publications include tables for selection
of joists and girders based on
load criteria. Provides specifications for design and installation.

Notes: _____

_____

TRIM

Architectural Woodwork Institute
P.O. Box 1550
13924 Braddock Rd., Suite 100
Centreville, VA 22020

Members consist of suppliers and manufacturers woodwork products. Provides technical data for architectural woodwork.

Notes: _____

_____

National Sash and Door Jobber Assoc.
10225 Robert Trent Jones Pky.
New Port Richey, FL 34655
(813) 372-3665
Fax (813) 372-2879
www.nsdjn.com

Members consist of wholesale distributors of windows, doors, millwork and related products. Provides technical information on millwork products.

Notes: _____

_____

Wood Moulding and Millwork
Producers Association
P.O. Box 25278
Portland, OR 97225
(503) 292-9288
Fax (503) 292-3490

Millwork and moulding is used as base trim, window trim and door casings. It is also used for chair rails, mantels and crown moulding. Members of this organization consist of manufacturers of millwork and wood moulding.

Notes: _____

_____

MISC.

American Architectural Manufacturers Assoc.
1540 E. Dundee Fd.
Palatine, IL 60067
(847) 202-1350
Fax (847) 202-1480

Members consist of manufacturers of such products as windows, storm doors, siding, gutters, downspouts, soffits, fascia, skylights, and sliding doors.

Notes: _____

_____

American Congress on Surveying and Mapping
5410 Grosvenor Lane, Suite 100
Bethesda, MD 20814
(301) 493-0200
www.survmap.org

Provides publications for the surveying and mapping industries. Monitors surveying regulations.

Notes: _____

_____

American Traffic Safety Services Assoc.
5440 Jefferson Davis Highway
Fredericksburg, VA 22407
(703) 898-5400
www.atssa.com

Provides manuals on uniform traffic control devices and work zone standards.

Notes: _____

_____

American Welding Society
550 LeJeune Rd., NW
Miami, FL 33126
(305) 443-9353
www.amweld.org

Provide technical information, approved welding techniques, codes, standards and specifications. Has over 100 committees that deal with individual topics of various aspects of welding.

Notes: _____

_____

Assoc. of the Wall and Ceiling Industry
307 E. Annandale Rd., Suite 200
Falls Church, VA 22042

Members composed of firms involved with installation of material used in ceiling and floors including drywall, plaster, insulation, fireproofing and acoustical tiles.

Notes: _____

_____

National Association of Home Builders
1201 15th St., NW
Washington, DC 20005
(800) 368-5242

Provides information on the technical, legal, and business aspects for homebuilders and others involved in the industry.

Notes: _____

_____

International Window Firm Assoc.
P.O. Box 42033
Scottsdale, AR 85080
(602) 595-9758
Fax (602) 595-9768

Provides information on tinting of all types of windows including windows in residential construction.

Notes: _____

_____

National Association of the Remodeling Industry
4900 Seminary Rd, Suite 320
Alexandria, VA 22311
(703) 575-1100
www.nari.org

Members consist of manufacturers, contractors, and distributors of products used in remodeling. Provides information on promotional aspects of remodeling industry.

Notes: _____

_____

National Fire Protection Assoc.
Robin Hill Corporate Park
Rte. 22, Box 1000
Patterson, NY 12563
(914) 878-4200
www.nfpa.org

Educational programs for
automatic fire protection
systems. Information on
building code changes and
regulations.

Notes: _____

_____

**National Organization to
Insure on Sound-Controlled
Environment**
1620 Eye St.
Washington, D.C. 20005
(202) 682-3901

Members are from government
and other organizations interested in reducing the noise
caused by airplane jets. Provides information and data on
abatement methods.

Notes: _____

_____

Scaffold Industry Association
14039 Sherman Way
Van Nuys, CA 91405
(818) 782-2012
www.scaffold.org

Provides information on devices
for supporting workers or
materials. Grading rules for
scaffolding planks and OSHA
standards available.

Notes: _____

_____

Vinyl Siding Institute
355 Lexington Ave., 11th Floor
New York, NY 10017
(212) 351-5400

Members consist of vinyl siding
manufacturers. Provides
information on installing vinyl
siding.

Notes: _____

_____

Chapter 12
Resources (Suppliers)

TOPICS   Address and telephone numbers of suppliers of
typical products used in light construction.

TIPS & INFORMATION

- Suppliers are excellent sources of information and assistance in
  problem solving. The majority of suppliers listed in this chapter
  have full time staff dedicated to assisting contractors and consumers in the proper use of their product.

- Always check if the supplier has a toll free number. Also consider
  the time zone you are contacting to call during business hours.

- Start your telephone call with a request to speak to a technical
  representative. A discussion with a nontechnical sales representative may not be as productive.

- After receiving a response to your inquiry, request a fax of any
  technical literature they have on the subject. This will allow you
  to confirm the opinions you were provided with as well as giving
  you material for future reference.

• Request contact information of local representative of the product. Local representatives may be able to actively assist in solving problems.

• Most of the technical staff for a supplier can assist in providing information. However, sometimes it is more time efficient when placing a followup call to speak to the same representative. Therefore, keep notes of the full name and also the extension number of the person you talked to.

• If a supplier cannot provide the requested information, ask the representative for a referral to a consultant or organization that can assist you.

• Many suppliers provide seminars on installation of their products. In fact, some suppliers certify installers. Being a certified installer is not only a good marketing strategy, it lets you fine tune your installation knowledge.

• Ask the supplier to send you an installation and/or a trouble, shooting guide. Valuable information about the product is contained in these publications.

BRICKS

The Belden Brick Co.
700 W. Tuscarawas
P.O. Box 20910
Canton, OH 44701
(330) 456-0031
Fax (330) 456-2694

Notes: _____

_____

ProSoCo, Inc.
755 Minnesota Ave.
P.O. Box 171677
Kansas City, KS 66117
(800) 255-4255
Fax (913) 281-2593

Notes: _____

_____

Blue Circle Cement
2 Pkwy. Center, Suite 1200
1800 Parkway Center
Marietta, GA 30067
(770) 423-4700
Fax (770) 423-4738

Notes: _____

_____

Richtex Corp.
P.O. Box 3307
Rickyard Rd.
Columbia, SC 29230
(803) 786-1260
Fax (803) 786-9703

Notes: _____

_____

Brady Brick & Supply Co.
1470 Abbott Dr.
Elgin, IL 60123
(708) 741-8343
Fax (708) 741-2262

Notes: _____

_____

Wisconsin Brick & Block Corp.
6399 Nesbit Rd.
Madison, WI 53744
(608) 845-8636
Fax (608) 845-5704

Notes: _____

_____

CONCRETE PRODUCTS

ITW Ramset/Red Head
Div. ITW
1300 N. Michael Dr.
Wood Dale, IL 60191
(630) 350-0370
Fax (630) 350-7985

Notes: _____

_____

Symons Corp.
100 E. Touhy Ave.
Des Plaines, IL 60018
(847) 298-3200
Fax (847) 635-9287

Notes: _____

_____

Lehigh Portland Cement Co.
Specialty Products Div.
7660 Imperial Way
Allentown, PA 18195
(800) 523-5488
Fax (610) 366-4638

Notes: _____

_____

Tamms Industries
3835 State Route 72
Kirkland, IL 60146
(800) 862-2667
Fax (815) 522-2323

Notes: _____

_____

Masonite Corp.
One South Wacker Dr.
Chicago, IL 60606
(800) 257-7885
Fax (312) 263-2850

Notes: _____

_____

U.S. Anchor Corp.
450 E. Copans Rd.
Pompano Beach, FL 33064
(800) 872-3330
Fax (800) 362-3320

Notes: _____

_____

DOOR HARDWARE

Baldwin Hardware
841 E. Wyomissing Blvd.
Reading, PA 19612
(610) 777-7811
Fax (610) 777-7256

Notes: _____

_____

Schlage Lock
1915 Jamboree
Colorado Springs, CO 80920
(800) 847-1864
Fax (415) 330-5627
www.schlagelock.com

Notes: _____

_____

Kwikset
1 Park Plaza, Suite 1000
Irvine, CA 92614
(714) 474-8800
Fax (714) 474-8879

Notes: _____

_____

Weiser
6700 Weisner Lock Dr.
Tucson, AZ 85746
(800) 677-LOCK
Fax (800) 688-LOCK

Notes: _____

_____

Precision Hardware
38100 Jaykay Dr.
Romulus, MI 48174
(313) 326-7500
Fax (313) 326-7540

Notes: _____

_____

Yale
P.O. Box 25288
Charlotte, NC 28229
(800) 438-1951
Fax (800) 338-0965
www.yalesecurity.com

Notes: _____

_____

ENGINEERED WOOD PRODUCTS

Alpine Engineered Products
P.O. Box 2225
Pompano Beach, FL 33061
(954) 781-3333
Fax (954)784-1647

Notes: _____

_____

Morton Buildings
252 W. Adams
Morton, IL 61550
(309) 263-7474
Fax (309) 266-5123

Notes: _____

_____

Georgia Pacific
133 Peachtree St.
Atlanta, GA 30303
(800) 284-5347
Fax (404) 230-5624

Notes: _____

_____

Trus Joist Macmillan
8644 154th Ave., N.E.
Boise, ID 83707
(208) 364-3650
Fax (208) 364-3633

Notes: _____

_____

Louisiana-Pacific Corp.
111 SW 5th Ave., Suite 4200
Portland, OR 97204
(800) 999-9105
(503) 796-0107

Notes: _____

_____

GARAGE DOORS

Amarr Garage Doors
5931 Grassy Creek
Winston Salem, NC 27105
(800) 503-3667
Fax (910) 744-0895

Notes: _____

_____

Delden Mfg. Co., Inc.
2130 Campbell St.
Kansas City, MO 64108
(800) 821-3708
Fax (816) 471-0943

Notes: _____

_____

Chamberlain Group, Inc.
Lift-Master Professional Products
845 Larch Ave.
Elmhurst, IL 60126
(800) 323-2276
Fax (630) 530-6091

Notes: _____

_____

General American Door Co.
(GADCO)
5050 Baseline Rd.
Montgomery, IL 60538
(630) 859-3000
Fax (630) 859-8122

Notes: _____

_____

Clopay Building Pro
Div. of Clopay Corp.
312 Walnut St., Suite 222
Cincinnati, OH 45202
(800) 282-2260
Fax (513) 762- 3519

Notes: _____

_____

Ideal Door
Div. Clopay
312 Walnut St.
Cincinnati, OH 45202
(513) 762-3566
Fax (513) 762-3519

Notes: _____

_____

FLOORING

American Marazzi Tile, Inc.
359 Clay Rd.
Sunnyvale, TX 75182
(972) 226-0110
Fax (972) 226-2263

Notes: _____

_____

Armstrong
P.O. Box 979
Guthrie, OK 73044
(405) 282-7584
Fax (405) 282-1130

Notes: _____

_____

Bruce Hardwood
16803 Dallas Parkway
Dallas, TX 75248
(800) 722-4647
Fax (214) 887-2110

Notes: _____

_____

Congoleum
3705 Quakerbridge Rd.
Mercerville, NJ 08619
(609) 584-3000
Fax (609) 584-3518

Notes: _____

_____

Pergo Flooring
P.O. Box 1775
Horsham, PA 19044
(800) 337-3746

Notes: _____

_____

Quarry Tile Co.
6328 E. Utah
Spokane, WA 99212
(509) 536-2812
Fax (509) 536-4072

Notes: _____

_____

GLASS

Cardinal IG
12301 Whitewater Dr.
Minnetonka, MN 55343
(612) 935-1722
Fax (612) 935-5538

Notes: _____

_____

PPG Industries
125 Colfax St.
Springdale, PA 15144
(800) 258-6398
Fax (412) 274-3837

Notes: _____

_____

Dow Corning
2200 W. Salzburg Rd.
Midland, MI 48686
(517) 496-4000
Fax (517) 496-4586

Notes: _____

_____

Therma-Tru
P.O. Box 8780
Maumee, OH 43537
(800) 537-8827
Fax (800) 322-8688

Notes: _____

_____

North American Glass
1001 Foster Ave.
Bensenville, IL 60106
(800) 323-2290
Fax (630) 595-3782

Notes: _____

_____

GYPSUM BOARD

Amico
3245 Fayette Ave.
Birmington, AL 35208
(800) 366-2642
Fax (205) 786-6527

Notes: _____

_____

Temple
P.O. Box N
Diboll, TX 75941
(409) 829-1485

Notes: _____

_____

Florida Stucco Products
21195 Boca Rio Rd.
Boca Raton, FL 33433
(561) 487-1301
Fax (561) 487-8536

Notes: _____

_____

Trim-Tex, Inc.
3700 W. Pratt Ave.
Lincolnwood, IL 60645
(800) 847-2333
Fax (847) 679-3017

Notes: _____

_____

National Gypsum
2001 Rexford Rd.
Charlotte, NC 28211
(704) 365-7300
Fax (704) 365-7222

Notes: _____

_____

U.S. Gypsum
125 S. Franklin St.
Chicago, IL 60606
(800) 874-4968
Fax (312) 606-5566

Notes: _____

_____

INSULATION

Apache Products Co.
107 Service Rd.
Anderson, SC 29625
(864) 964-2720
Fax (864) 964-2721

Notes: _____

_____

Dow Chemical Co.,
2020 Dow Center
P.O. Box 1206
Midland, MI 48674
(800) 441-4369
Fax (517) 832-1465

Notes: _____

_____

Celotex Corp.
P.O. Box 31602
Tampa, FL 33631
(813) 873-1700
Fax (813) 873-4413

Notes: _____

_____

Johns Manville Corp.
717 17th St.
Denver, CO 80202
(800) 654-3103
Fax (303) 978-2318

Notes: _____

_____

CertainTeed Corp.
750 E. Swedesford Rd.
P.O. Box 860
Valley Forge, PA 19482
(800) 233-8990
Fax (610) 341-7940

Notes: _____

_____

Tenneco Building Products
2907 Log Cabin Dr.
Smyrna, GA 30080
(800) 241-4402
Fax (404) 350-1489

Notes: _____

_____

MILL WORK

Aged Woods, Inc.
2331 E. Market St.
York, PA 17402
(800) 233-9307
Fax (717) 840-1468

Notes: _____

_____

Pecora Corp.
165 Wambold Rd.
Harleysville, PA 19438
(800) 523-6688
Fax (215) 721-0286

Notes: _____

_____

Custom Decorative Mouldings
P.O. Box F
Greenwood, DE 19950
(800) 543-0553
Fax (302) 349-4816

Notes: _____

_____

Superior Moulding
5953 Sepulveda Blvd.
Van Nuys, CA 91411
(818) 376-1415
Fax (818) 376-1314

Notes: _____

_____

Ornamental Mouldings
3804 Comanche Rd.
Archdale, NC 27263
(800) 779-1135
Fax (910) 431-9104

Notes: _____

_____

Turncraft
P.O. Box 2429
White City, OR 97504
(541) 826-2911
Fax (541) 826-1393

Notes: _____

_____

PAVER BLOCKS

Buechel Stone Corp.
W3639 Highway H
Chilton, WI 53014
(800) 236-4473
Fax (414) 849-7810

Notes: _____

_____

Uni-Group U.S.A.
4362 Northlake Blvd., Ste 20
Palm Beach Gardens, FL 33410
(800) 872-1864
Fax (561) 627-6403

Notes: _____

_____

Keystone Retaining Wall
Systems, Inc.
4444 W. 78th St.
Minneapolis, MN 55435
(800) 891-9791
Fax (612) 897-3858

Notes: _____

_____

Unilock
International Blvd.
Brewster, NY 10509
(800) 864-5625
Fax (914) 278-6788

Notes: _____

_____

PAINTS

Cabot Stains
100 Hale St.
P.O. Box 807
Newburport, MA 01950
(508) 465-1900
Fax (508) 462-0511

Notes: _____

_____

Pratt & Lambert Paints
101 Prospect Ave.
Cleveland, OH 44115
(800) 289-7728
Fax (716) 877-9646

Notes: _____

_____

**DuPont Co., High
Performance Coatings**
P.O. Box 80021 BMP21-1308
Wilmington, DE 19880
(302) 992-4928
Fax (902) 892-5695

Notes: _____

_____

PPG Industries, Inc.
125 Colfax St.
Springdale, PA 15144
(800) 258-6398
Fax (412) 274-3837

Notes: _____

_____

Duron Paints & Wallcoverings
10406 Tucker St.
Beltsville, MD 20705
(800) 723-8766
Fax (301) 595-0429
Regional: M, S, N

Notes: _____

_____

Sherwin Williams Co.
Stores Division
101 Prospect Ave.
10 Midland Bldg.
Cleveland, OH 44115
(216) 566-2000
Fax (216) 566-1392

Notes: _____

_____

PLUMBING FIXTURES

Crane Plumbing/Fiat Products
1235 Hartrey Ave.
Evanston, IL 60202
(847) 864-9777
Fax (847) 864-7652

Notes: _____

_____

Kohler Co.
Plumbing Div.
444 Highland Dr.
Kohler, WI 53044
(414) 457-4441
Fax (414) 457-6952

Notes: _____

_____

Delta Faucet Co.
55 E. 111th St.
Indianapolis, IN 462??
(317) 848-1812
Fax (317) 573-3492

Notes: _____

_____

Moen Inc.
25300 Al Moen Dr.
North Olmsted, OH 44070
(800) 289-6636
Fax (800) 628-0894

Notes: _____

_____

Gerber Plumbing Fixtures Corp.
800 W. Touhy Ave.
Chicago, IL 60646
(847) 675-6570
Fax (800) 543-7237

Notes: _____

_____

Sterling Plumbing Group
2900 W. Golf Rd.
Rolling Meadows, IL 60008
(847) 734-1777
Fax (847) 734-4767

Notes: _____

_____

ROOF DECKING

ATAS International, Inc.
6612 Snowdrift Rd.
Allentown, PA 18106
(610) 395-8445
Fax (610) 395-9342

Notes: _____

_____

Petersen Metal Products
2301 Success Dr.
Odessa, FL 33556
(813) 372-1100
Fax (813) 372-8112

Notes: _____

_____

Butler Mfg. Co.
BMA Tower, Penn Valley Park
Kansas City, MO 64141
(816) 968-3525
Fax (819) 968-3720

Notes: _____

_____

VP Buildings
1000 Poplar, Suite 400
Memphis, TN 38119
(901) 767-5910
Fax (910) 762-6010

Notes: _____

_____

McElroy Metal, Inc.
1500 Hamilton Rd.
Bossier City, LA 71111
(800) 950-6531
Fax (318) 747-8029

Notes: _____

_____

SEPTIC TANKS

American Concrete Products, Inc.
6859 Q St.
Omaha, NE 68117
(402) 331-5775
Fax (402) 331-0742
Regional: M

Notes: _____

_____

Western Concrete Products Co., Inc.
510 5th St.
Cadillac, MI 49601
(616) 775-3466
(616) 775-1848

Notes: _____

_____

Baystar Precast Corp.
925 Skinners Turn Rd.
Owings, MD 20736
(410) 257-6777
Fax (410) 465-5540

Notes: _____

_____

SHINGLE ROOFING

CertainTeed Corp.
750 E. Swedesford Rd.
P.O. Box 860
Valley Forge, PA 19482
(800) 233-8990
Fax (610) 341-7940

Notes: _____

_____

GS Roofing Products, Inc.
5525 Mac Arthur Blvd.
Suite 900
Irving, TX 75038
(972) 580-5604
Fax (972) 580-5692

Notes: _____

_____

Elk Corp.
14643 Dallas Parkway
Suite 1000
Dallas, TX 75240
(800) 354-7732
Fax (972) 851-0447

Notes: _____

_____

3M Specified Construction
Products Dept.
3M Center Bldg. 225-4S-08
St. Paul, MN 55144
(800) 480-1704
Fax (612) 736-0611

Notes: _____

_____

GAF Materials Corp.
1361 Alps Rd.
Wayne, NJ 07470
(800) 766-3411
Fax (201) 628-3356

Notes: _____

_____

SIDING

Alcoa Building Products
1501 Michigan, P.O. Box 716
Sidney, OH 45365
(800) 962-6937
Fax (937) 498-6396

Notes: _____

_____

Masonite Corp.
One South Wacker DR.
Chicago, IL 60606
(800) 257-7885
Fax (312) 263-2850

Notes: _____

_____

CertainTeed Corp.
750 E. Swedesford Rd.
P.O. Box 860
Valley Forge, PA 19482
(800) 233-8990
Fax (610) 341-7940

Notes: _____

_____

Rollex Corp.
2001 Lunt Ave.
Elk Grove Village, IL 60007
(800) 251-3300
Fax (800) 521-8965

Notes: _____

_____

Georgia-Pacific Corp.
133 Peachtree St.
Atlanta, GA 30303
(800) 284-5347
Fax (404) 230-5624

Notes: _____

_____

Wolverine Vinyl Siding
750 E. Swedesford Rd.
Valley Forge, PA 19482
(888) 838-8100
Fax (610) 341-7538

Notes: _____

_____

WINDOWS

Anderson Windows
100 Fourth Ave., North
Bayport, MN 55003
(800) 426-7691
Fax (512) 434-1542
www.andersonwindows.com

Notes: _____

_____

Caradon Doors and Windows
4350 Peachtree Industrial Blvd.
Norcross, GA 30091
(770) 497-2000

Notes: _____

_____

CertainTeed Corp.
750 E. Swedesford Rd.
P.O. Box 860
Valley Forge, PA 19482
(800) 233-8990
Fax (610) 341-7940

Notes: _____

_____

Hurd Millwork Company
575 S. Whelen Ave.
Medford, IL 54451 ???
(800) 223-4873
Fax (715) 748-6043
www.hurd.com

Notes: _____

_____

Pella Corporation
102 Main St.
Pella, IA 50219
(515) 628-1000
Fax (515) 628-6457

Notes: _____

_____

Weather Shield Inc.
One Weather Shield Plaza
P.O. Box 309
Medford, WI 54451
(800) 477-6808
(715) 748-6999

Notes: _____

_____

WATERPROOFING & DAMPPROOFING

American Hydrotech, Inc.
303 E. Ohio St.
Chicago, IL 60611
(800) 877-6125
Fax (312) 661-0731

Notes: _____

_____

Mar-Flex Systems, Inc.
6866 Chrisman Lane
Middletown, OH 45042
(800) 498-1411
Fax (513) 422-7282

Notes: _____

_____

Carlisle Coatings & Water-proofing, Inc.
8810 W. 100th St., S
Sapulpa, OK 74067
(800) 338-8701
Fax (918) 224-2030

Notes: _____

_____

TC Mira DRI
3500 Parkway Lane
Norcross, GA 30092
(770) 447-627?
Fax (770) 729-182?

Notes: _____

_____

Grace Construction Products
62 Whittemore Ave.
Cambridge, MA 02140
(617) 498-4994
Fax (617) 498-4994

Notes: _____

_____

WOOD TREATMENTS

Chemical Specialities, Inc.
200 E. Woodlawn Rd.
Suite 250
Charlotte, NC 28217
(800) 421-8661
Fax (704) 527-8232

Notes: _____

_____

Hickson Corp.
1955 Lake Park Dr.
Suite 250
Smyrna, GA 30080
(770) 801-6600
Fax (770) 801-1990

Notes: _____

_____

Thompson Co.
Div. The Thompson
Miniwax Co.
825 Crossover Lane
Memphis, TN 38117
(901) 685-7555
Fax (901) 763-2096

Notes: _____

_____

Index

About the Author

Dr. August W. Domel, Jr., is a licensed structural engineer, professional engineer and attorney at law in Illinois. Currently he is Adjunct Assistant Professor at the Illinois Institute of Technology (I.I.T.), Judson College, and manager of structural engineering at Engineering Systems, Inc. in Aurora, Illinois.

He graduated Summa Cum Laude from Bradley University with a bachelor of science degree in civil engineering. He received a master's degree and Ph.D. in civil engineering from I.I.T and the University of Illinois at Chicago, respectively. He also earned a law degree from Loyola University of Chicago.

He has written books on the topics of earthquake design of high-rise buildings, design of water-retaining structures, concrete floor design and estimating. He lives with his wife and five children in Sleepy Hollow, Illinois